SYNGRESS®

W9-BEO-656

WIRELESS HACKING

Projects for Wi-Fi Enthusiasts

By the SoCalFreeNet.org Wireless Users Group
Lee Barken *with*
Eric Bermel, John Eder, Matthew Fanady
Michael Mee, Marc Palumbo, Alan Koebrick

KEY	SERIAL NUMBER
001	HJCV184764
002	PO5FGHJ887
003	82JH26765V
004	VBHF43299M
005	C23NMVCXZ3
006	VB5T883E4F
007	HJJ3EBNBB6
008	2987GMKKMM
009	629JT5678N
010	IMWT6T3456

PUBLISHED BY
Syngress Publishing, Inc.
800 Hingham Street
Rockland, MA 02370

Wireless Hacking: Projects for Wi-Fi Enthusiasts

Printed in the United States of America
1 2 3 4 5 6 7 8 9 0
ISBN: 1-931836-37-X

Publisher: Andrew Williams   Page Layout and Art: Patricia Lupien
Acquisitions Editor: Christine Kloiber   Copy Editor: Mike McGee
Technical Editor: Lee Barken   Indexer: Odessa&Cie
Cover Designer: Michael Kavish

Distributed by O'Reilly Media, Inc. in the United States and Canada.
For information on rights and translations, contact Matt Pedersen, Director of Sales and Rights, at Syngress Publishing; email matt@syngress.com or fax to 781-681-3585.

Register for Free Membership to

solutions@syngress.com

Over the last few years, Syngress has published many best-selling and critically acclaimed books, including Tom Shinder's *Configuring ISA Server 2000*, Brian Caswell and Jay Beale's *Snort 2.0 Intrusion Detection*, and Angela Orebaugh and Gilbert Ramirez's *Ethereal Packet Sniffing*. One of the reasons for the success of these books has been our unique **solutions@syngress.com** program. Through this site, we've been able to provide readers a real time extension to the printed book.

As a registered owner of this book, you will qualify for free access to our members-only solutions@syngress.com program. Once you have registered, you will enjoy several benefits, including:

- Four downloadable e-booklets on topics related to the book. Each booklet is approximately 20-30 pages in Adobe PDF format. They have been selected by our editors from other best-selling Syngress books as providing topic coverage that is directly related to the coverage in this book.

- A comprehensive FAQ page that consolidates all of the key points of this book into an easy to search web page, providing you with the concise, easy to access data you need to perform your job.

- A "From the Author" Forum that allows the authors of this book to post timely updates links to related sites, or additional topic coverage that may have been requested by readers.

Just visit us at **www.syngress.com/solutions** and follow the simple registration process. You will need to have this book with you when you register.

Thank you for giving us the opportunity to serve your needs. And be sure to let us know if there is anything else we can do to make your job easier.

SYNGRESS®

Acknowledgments

Syngress would like to acknowledge the following people for their kindness and support in making this book possible.

Syngress books are now distributed in the United States and Canada by O'Reilly Media, Inc. The enthusiasm and work ethic at O'Reilly is incredible and we would like to thank everyone there for their time and efforts to bring Syngress books to market: Tim O'Reilly, Laura Baldwin, Mark Brokering, Mike Leonard, Donna Selenko, Bonnie Sheehan, Cindy Davis, Grant Kikkert, Opol Matsutaro, Steve Hazelwood, Mark Wilson, Rick Brown, Leslie Becker, Jill Lothrop, Tim Hinton, Kyle Hart, Sara Winge, C. J. Rayhill, Peter Pardo, Leslie Crandell, Valerie Dow, Regina Aggio, Pascal Honscher, Preston Paull, Susan Thompson, Bruce Stewart, Laura Schmier, Sue Willing, Mark Jacobsen, Betsy Waliszewski, Dawn Mann, Kathryn Barrett, John Chodacki, and Rob Bullington.

The incredibly hard working team at Elsevier Science, including Jonathan Bunkell, Ian Seager, Duncan Enright, David Burton, Rosanna Ramacciotti, Robert Fairbrother, Miguel Sanchez, Klaus Beran, Emma Wyatt, Rosie Moss, Chris Hossack, Mark Hunt, and Krista Leppiko, for making certain that our vision remains worldwide in scope.

David Buckland, Marie Chieng, Lucy Chong, Leslie Lim, Audrey Gan, Pang Ai Hua, and Joseph Chan of STP Distributors for the enthusiasm with which they receive our books.

Kwon Sung June at Acorn Publishing for his support.

David Scott, Tricia Wilden, Marilla Burgess, Annette Scott, Andrew Swaffer, Stephen O'Donoghue, Bec Lowe, and Mark Langley of Woodslane for distributing our books throughout Australia, New Zealand, Papua New Guinea, Fiji Tonga, Solomon Islands, and the Cook Islands.

Winston Lim of Global Publishing for his help and support with distribution of Syngress books in the Philippines.

Technical Editor & Contributor

Lee Barken CISSP, CCNA, MCP, CPA, is the co-director of the Strategic Technologies And Research (STAR) Center at San Diego State University (SDSU) and the President and co-founder of SoCalFreeNet.org, a non-profit community group dedicated to building public wireless networks. Prior to SDSU, he worked as an IT consultant and network security specialist for Ernst & Young's Information Technology Risk Management (ITRM) practice and KPMG's Risk and Advisory Services (RAS) practice. Lee is the technical editor for *Mobile Business Advisor Magazine*, and writes and speaks on the topic of wireless LAN technology and security. He is the author of *How Secure Is Your Wireless Network? Safeguarding Your Wi-Fi LAN* (ISBN 0131402064) and co-author of *Hardware Hacking: Have Fun While Voiding Your Warranty* (ISBN 1932266836).

Lee is the author of Chapter 1 "A Brief Overview of the Wireless World," Chapter 2 "SoCalFreeNet.org: An Example of Building Large Scale Community Wireless Networks," Chapter 4 "Wireless Access Points," Chapter 8 "Low-Cost Commercial Options," and Appendix A "Wireless 802.11 Hacks."

"The most precious possession that ever comes to a man in this world is a woman's heart."

—Josiah G. Holland

To the love of my life, Stephanie:
Thank you for your never-ending love and encouragement.

Contributors

Eric Bermel is an RF Engineer and Deployment Specialist. He has many years of experience working for companies such as Graviton, Western US, Breezecom, Alvarion, and PCSI. Eric has extensive experience developing and implementing RF site surveys, installation and optimization plans for indoor and outdoor ISM and U-NII band systems.

Eric is the author of Chapter 10 "Antennas."

John Eder (CISSP, CCNA) is a security expert with Experian. He currently provides strategic and technical consulting on security policy and implementation. His specialties involve: risk profiling, wireless security, network security, encryption technologies, metrics development and deployment, and risk analysis. John's background includes a position as a consultant in the Systems and Technology Services (STS) practice at Ernst & Young, LLP.

John holds a bachelor's degree from San Diego State University. He actively participates in the security community, making presentations and writing numerous articles on wireless security. John is a proud member of SoCalFreeNet.

John enjoys the support of his loving wife Lynda, a caring family (Gabriel, Lyn, and Genevieve), and a great friendship with his director, Michael Kurihara. The security information in this book was made possible through the help of the m0n0wall team, the Soekris Engineering team, the West Sonoma County Internet Cooperative Corporation, and the many members of SoCalFreeNet.

John is the author of Chapter 3 "Securing Our Wireless Community."

Matthew Fanady is a gear-head turned networking and computer enthusiast, and has been wrenching on cars and building computers since he was 16 years old. He is currently employed designing and constructing electric vehicles for a small startup company in San Diego, and spends his free time troubleshooting computers and exploring new ways to incorporate the latest communications technologies into everyday life. Matthew was one of the early pioneers of community wireless networks. In 2002, he began building a grass-roots community wireless network in his own neighborhood of Ocean Beach, where he was able to bring his passion for hacking together with his passion for wrenching. His efforts, along with those of others in San Diego, led to the inception of SoCalFreeNet which continues to build community-based wireless networks in San Diego.

Matthew is the author of Chapter 11 "Building Outdoor Enclosures and Antenna Masts," and Chapter 12 "Solar-Powered Access Points and Repeaters."

Alan Koebrick is the Vice President of Operations for SoCalFreeNet.org. He is also a Business Systems Analyst with a large telecommunications company in San Diego. Alan has a Bachelors degree in E-Business from the University of Phoenix. Prior to his current job, Alan spent 4 years with the United States Marine Corps where he performed tasks as a Network Administrator and Legal Administrative Clerk. Alan is also launching a new venture, North County Systems, a technology integrator for the Small Office / Home Office market.

Alan is the author of Chapter 5 "Wireless Client Access Devices."

Michael Mee Michael started building his own computers after discovering the TRS-80 at Radio Shack years ago. He went on to work for a software startup, before dot coms made it fashionable. Then he had several great years at Microsoft, back when 'the evil empire' meant IBM. There he worked on database products like Access and Foxpro for Windows. Returning to his hacking roots, he's now helping build high-speed community wireless for users everywhere, especially through SoCalFreeNet.org.

Michael is the author of Chapter 6 "Wireless Operating Systems," and Chapter 7 "Monitoring Your Network."

Marc Palumbo (Society of Mechanical Engineers #4094314) is the Creative Director for the SoCalFreeNet.org. He is an Artist/Engineer and the owner of Apogee Arts, headquartered in San Diego, California. His company builds Community Networks, provisions Internet access for business and residential use, and designs and executes LANS purposed for specific vertical markets such as graphics, video editing, publishing, and FDA regulated manufacturing. He has built secure wireless surveillance systems deployed in Baghdad, Iraq, and for Homeland Security. Noteworthy wireless triage installations include the city of Telluride,

Colorado, and Black Rock Desert, Nevada for Burning Man. Marc holds a bachelors degree from the University of Maryland, received a National Endowment for the Arts stipend, and was a Fellow at the Center for Advanced Visual Studies, MIT. He began building his first computers in 1978 as part of his voice activated pyrotechnic interactive sculpture, "Clytemnestra." The work won a once in 20-year honor for the Boston Arts Festival, 1985. He built his first RF device to light high voltage Neon works of art.

Marc also helped deliver the first paint package for the PC, Splash! with Spinnaker Software and LCS Telegraphics. He created the first digital images for the PC, and his digital imagery has been published in Smithsonian Magazine, Volume 11, Number 9, Dec. 1980, pp. 128-137 and Macworld Magazine, October 1988, pp. 96 through April 1989. One of the first Artists to use lasers for art, he has created large-scale images in the sky, mountains, and in the urban landscape. He has worked for and appeared on national television, including "Race for the High Ground", Frontline News with Jessica Savitch (S.D.I. Demo of Star Wars Defense System, laser destroying satellite, W.G.B.H., Boston, MA, April 1983). He has also worked on production and on air talent crews for Discover Magazine's TV show with James (Amazing) Randi, "A Skeptic's Guide" March 1999.

Working with Miami Springs High School and his corporate sponsor, Symbiosis, he created a team to build a robot to compete in Dean Kamen's US First Competition, a program to encourage engineering careers for high school students.

Marc is the author of Chapter 9 "Mesh Networking."

Foreword Contributor

Rob Flickenger has been hacking systems all of his life, and has been doing so professionally for over ten years. He is one of the inventors of NoCat, and is also an active member of FreeNetworks.org. Rob has written and edited a number of books for O'Reilly & Associates, including *Wireless Hacks* and *Building Wireless Community Networks*. He is currently a partner at Metrix Communication LLC in Seattle, WA (http://metrix.net/).

Contents

Foreword

Over the last couple of years, manufacturers have produced some incredibly sophisticated wireless networking equipment. Consumers' ongoing demand for low-cost, high-speed, easy-to-use networking gear has forced hardware manufacturers to continually refine their designs. Engineers have produced tiny devices that use very little power to perform amazing feats of ingenuity, producing them on such a large scale that the cost is staggeringly low.

Unfortunately, these wireless gadgets nearly always have one real drawback: they are designed to appeal to the widest possible market. Out of the box, they will do what marketing folks think people want, and not much more. One particular radio card may be small and common, but it doesn't have very good range. An access point might be easy to use, but it doesn't support many clients. Another may work well, but only with Windows. And of course, it's nearly impossible to learn any of this from the printing on the box. This is perfectly understandable, since some corners nearly always need to be cut to bring down the overall cost of the design. Mass industry may be able to produce the perfect $10 toaster, but you wouldn't want to use it to cook a Thanksgiving dinner for eight.

But with people like Lee Barken, engineering never really stops; even after the gear is shrink wrapped in a pretty box and placed on a shelf in a store. Wireless hackers, like Lee and the other members of SoCal Free Net, are not afraid to rip open the pretty injection molded plastic facade and wrangle the absolute best possible performance out of a piece of networking gear. By maximizing the performance / cost trade-off of the equipment they use, they are bringing real, fast, free

Internet access to the masses. FreeNetworks folks can do amazing things with that $10 toaster, and can tell you definitively when and how to upgrade.

Wireless hackers are a strange breed of computer geek. They need not only understand the bits and protocols of networks and the frequencies and dBm of radio. Perhaps even more than technical skill, the would-be community wireless guru needs to be a master at networking with people. If the great community wireless networking experiment of the last four or five years has taught us anything, it's that no project can succeed without the support and full participation of your local community. Knowing how to do it solves only half of the problem; getting the job done through an efficient and sustainable organization is just as important.

This book will get you into the mindset of a wireless hacker. You will learn about what the current technology is really capable of providing, as well as extending it to provide more than you likely thought possible. You'll understand the social, performance, and security trade-offs involved in choosing a network topology. You will learn about cutting edge technologies like mesh and solar powered gear. And just as importantly, you will learn the story of how SoCal Free Net is organizing to build community wireless networks on a large scale in southern California.

Today, there are hundreds of community wireless network projects being built all over the globe. Cities like Austin, New York, Seattle, San Francisco, and Portland have created huge public wireless networks that are entirely sustained by local people who want to have a stake in their communications network. Like any public works project, community networks bring important benefits to the people who use them, specifically because they are open and free to be used by anyone. Take what you learn from this book, and use it to turn your own community into a well connected wireless utopia.

—Rob Flickenger

Introduction

Dear Readers,

Thank you for adding this book to your library of wireless information. Our goal is to educate, inspire and be a resource for building your own community wireless networks. If you're like me, you have an office (or garage) filled with wireless equipment and toys. Now it's time to put that gear to work! With this book as your guide, you can harness the power of wireless communications to (legally) share your bandwidth with your neighbors, help provide broadband connectivity to low-income underprivileged neighborhoods, or just have fun playing around with Wi-Fi projects.

Always remember to involve your best friends in your Wi-Fi projects. My girlfriend, Stephanie, has always been incredibly supportive of all my wireless activities, even helping me on my strict Pringles Can diet to help us stock up on cans for building antennas. What makes Wi-Fi fun is that it is the first new technology to come around in awhile that actually encourages you to step away from your monitor, go outside and interact with your neighbors!

Will the next generation of wireless enthusiasts change the landscape of broadband Internet access? I sure think so. Using the knowledge and projects in this book, you can take your wireless network to the next level. Whether you're technically inclined, or just a novice, these practical tutorials and step-by-step project instructions will help you marry your sense of Wi-Fi adventure with your desire to help promote free wireless networks and enhance communities.

Groups across the country are joining together to harness the power of public wireless networks. In San Diego, we created SoCalFreeNet to bring together likeminded wireless enthusiasts with a special purpose. Our mission statement says it all:

> "SoCalFreeNet.org is a non-profit community group dedicated to building, deploying and growing public wireless networks to increase widespread broadband adoption and create an empowered, connected society, where technology becomes a community resource."

And we are strong believers in the saying, "Think Globally, Act Locally."

Now is the time to join a group in your area (or start one!) to help promote community wireless networks and the philosophy of free and open broadband access. Will you join me? If so, keep reading...

—*Lee Barken*
September 2004

Part I

Introduction to Wireless Hacking

Chapter 1

A Brief Overview of the Wireless World

Topics in this Chapter:

- Introduction to Wi-Fi
- The History and Basics of 802.11
- Why Wi-Fi?

Introduction to Wi-Fi

Welcome to the world of "wireless magic." 802.11 (Wi-Fi) wireless LANs have exploded onto the scene with an excitement not seen since the introduction of the Internet itself. Getting rid of the wires means getting rid of the hassle. With Wi-Fi, you can roam through your favorite coffee shop, boardroom or living room, all the while maintaining the convenience of high-speed connectivity. With Wi-Fi, life is good!

Once you've gone through the experience of setting up a local Wi-Fi network, your first instinct is likely to think that "bigger is better." Expanding the coverage and increasing the network cloud enables you to share Internet resources with others as a community service. As long as you're not breaking the law (or violating your ISP's terms of service), you should listen to that creative instinct and start getting together with your neighbors (or others in your community) and begin planning a community wireless network! This book is all about the challenges, thrills, and adventures of such an effort, started in San Diego, and known as "SoCalFreeNet." By offering this roadmap to you, the reader, we hope to inspire more such efforts in neighborhoods all around the world.

In this chapter, we will cover some of the Wi-Fi basics you'll need to understand in order to complete the wireless hacks found throughout this book. We start with the history and basics of 802.11, including a review of the differences between *a*, *b* and *g*. Next, we cover wireless architectures and FCC rules. Finally, we include some discussion on the social benefits of community wireless networking.

The History and Basics of 802.11

The desire of people to communicate wirelessly spans many generations and technologies. Some might even argue that the ancient activity of lighting fires and using smoke signals was an early attempt to distribute a message without wires. In this book, however, we refer to the term "wireless" in the context of a modern data network. In other words: the ability to transmit and receive binary data from one location to another. A great deal of wireless data technology evolved in the late 20th century. Unfortunately, these wireless devices were typically proprietary and expensive. Their uses included specialized applications, such as remote cash registers and warehouse inventory systems.

After spending the better part of the 1990s engaged in technical discussions, the Institute of Electrical and Electronics Engineers (IEEE) ratified the 802.11 protocol in 1997. The original protocol supported three physical layer definitions: Direct Sequence Spread Spectrum (DSSS), Frequency Hopping Spread Spectrum (FHSS), and InfraRed (IR). The supported data rates for DSSS and FHSS were 1 and 2 Mbps. These protocols operated in the 2.4 GHz unlicensed spectrum. IR remains an interesting footnote in the history of 802.11, as it never achieved any notable commercial success due to its limited range and line of sight requirements.

In 1999, the higher speed 802.11a and 802.11b protocols were ratified. 802.11b added 5.5 and 11 Mbps support using DSSS in 2.4 GHz, making it backwards-compatible with existing 1 and 2 Mbps DSSS gear (but not compatible with FHSS or IR equipment). 802.11a added Orthogonal Frequency Division Multiplexing (OFDM) as a modulation technique in the 5 GHz unlicensed spectrum, with speeds of up to 54 Mbps. In 2003, 802.11g was ratified, which provided higher speeds (up to 54 Mbps). 802.11g works by applying OFDM modulation techniques in the 2.4 GHz unlicensed spectrum. It remains backwards-compatible with 802.11b by integrating DSSS modulation (at 11, 5.5, 2, and 1 Mbps).

NOTE...WHAT IS THE WI-FI ALLIANCE?

If the IEEE publishes the 802.11 standards that define how wireless communication works, then why do we have the Wi-Fi Alliance? The answer is simple: Although standards do exist, the Wi-Fi Alliance performs interoperability testing and also serves as a non-profit, industry advocate. Once a product has successfully completed interoperability testing, it can use the "Wi-Fi" trademark on its packaging. Surprisingly, up to 25 percent of products do not pass the testing process on their first run! With the Wi-Fi seal on various products, consumers can rest assured that an Access Point purchased from one company will work with a PCMCIA card purchased from another company. This kind of guaranteed interoperability has been a major factor in the widespread adoption and prolific success of mass-produced 802.11 gear. For more information about the Wi-Fi Alliance, you can visit www.wi-fialliance.com.

IEEE Alphabet Soup

Understanding the differences between the various IEEE protocols can be daunting. Consumers often make the mistake of purchasing incompatible hardware and then returning it to the computer store when it doesn't work. In this section, we will clear up any confusion about the differences between 802.11a, b, and g. Using this information as a guide, you will be ready to make informed and educated choices regarding the protocol best suited for your particular deployment.

802.11b

For many years, 802.11b was widely regarded as the most popular form of Wi-Fi. It utilizes frequencies in the 2.4 GHz range (2.400–2.485GHz) and has 11 channels. However, only three of these channels are truly non-overlapping. See Table 1.1 for a list of all channels. The range (distance) for 802.11b can vary widely, but each access point (with default antennas) typically covers a few hundred feet (indoors) or a few thousand feet (outdoors). With specialized, external antennas, this range can be greatly increased. 802.11b operates in the Industrial, Scientific, and Medical (ISM) unlicensed spectrum. More about ISM in a moment.

Table 1.1 802.11b Channels

Channel Number	Center Frequency (in GHz)	USA	Europe	Spain	France	Israel	China	Japan
1	2.412	✓	✓				✓	✓
2	2.417	✓	✓				✓	✓
3	2.422	✓	✓			✓	✓	✓
4	2.427	✓	✓			✓	✓	✓
5	2.432	✓	✓			✓	✓	✓
6	2.437	✓	✓			✓	✓	✓
7	2.442	✓	✓			✓	✓	✓
8	2.447	✓	✓			✓	✓	✓
9	2.452	✓	✓			✓	✓	✓
10	2.457	✓	✓	✓	✓		✓	✓
11	2.462	✓	✓	✓	✓		✓	✓
12	2.467		✓		✓			✓
13	2.472		✓		✓			✓
14	2.484							✓

The top speed for 802.11b is 11 Mbps, but it will auto-negotiate down to rates of 5.5, 2, and 1 Mbps as the signal strength deteriorates. These speeds include a relatively high amount of "overhead," as required by the protocol to operate. Keep in mind that actual throughput (for all 802.11 flavors) is typically about 50–60 percent of the advertised speeds. In other words, even under ideal circumstances, the actual data throughput (say, transferring a file) is usually around a maximum of 5–6 Mbps.

NOTE...MEASURING THROUGHPUT

An excellent tool for measuring actual throughput is QCheck from Ixia (formerly from NetIQ). Best of all, the tool is free! You can download a copy at www.ixiacom.com/products/performance_applications/pa_display.php?skey=pa_q_check.

So many people have discovered the joys of wireless networking that 802.11b is quickly becoming a victim of its own success. Specifically, the level of Wi-Fi congestion found in any major metropolitan area is raising the RF noise floor and rendering many long distance links unusable. The pros and cons of 802.11b are as follows:

- Upside: Most popular and widely available; least expensive; good coverage
- Downside: Relatively slow speed; interference from other 2.4 GHz devices; only three non-overlapping channels

NEED TO KNOW...WHAT ARE NON-OVERLAPPING CHANNELS?

You may be wondering: If an 802.11b channel is 22 MHz wide and operates in a band of frequencies that is 83.5 MHz wide, then how do we fit in 11 channels? The answer is that we "cheat." Each center frequency is spaced only 5 MHz apart, meaning that each channel overlaps with other, adjacent channels. The only way to have truly non-overlapping (and non-interfering) channels is to use Channels 1, 6, and 11. That's why you sometimes hear people say that 802.11b only has three "real" channels.

802.11a

Although 802.11a was released around the same time as 802.11b, it never achieved the same critical mass or wide scale acceptance. This was despite 802.11a's superior 54-Mbps speed. (Before 802.11g was released, 802.11a was the fastest Wi-Fi protocol available.) 802.11a operates in the 5 GHz spectrum and has 12 non-overlapping channels. As a result of this higher frequency, 802.11a has a much harder time penetrating through obstacles, such as walls and other objects. This results in a range much lower than 802.11b.

One major advantage of 802.11a is that it is less prone to interference from other 5 GHz devices simply because there are fewer 802.11a and 5 GHz cordless devices deployed in the real world to compete with. Unlike 2.4 GHz (which is flooded with competing devices), 5 GHz remains relatively unused. This trend, however, is changing as more and more cordless phones and other gadgets are migrating to the less crowded 5 GHz spectrum. However, for the time being, 802.11a makes an excellent choice for building-to-building and backhaul solutions where line of sight is available. Also, 802.11 offers 11 non-overlapping channels. See Table 1.2 for a list of 802.11a channels. The pros and cons of 802.11a are as follows:

- Upside: Relatively fast speed; more non-overlapping channels than 802.11b/g; 5 GHz spectrum is less crowded

- Downside: More expensive; shorter range

Table 1.2 802.11a Channels

Channel Number	Center Frequency (in GHz)	USA	Europe (some countries may vary)	Japan	Taiwan	China
36	5.180	✓	✓	✓		
40	5.200	✓	✓	✓		
44	5.220	✓	✓	✓		
48	5.240	✓	✓	✓		
52	5.260	✓	✓		✓	
56	5.280	✓	✓		✓	
60	5.300	✓	✓		✓	
64	5.320	✓	✓		✓	
149	5.745	✓				✓
153	5.765	✓				✓
157	5.785	✓				✓
161	5.805	✓				✓

NEED TO KNOW…BACKHAULS

A "backhaul" is the name typically given to the mechanism for delivering bandwidth to the Access Point. For example, if your DSL/cable connection terminates in one building, but you'd like to deliver bandwidth and create a hotspot in another building a few miles away, then you could create a wireless link between the buildings and set up a second hotspot in the new building. The wireless link that connects the DSL/cable connection in the first building to the hotspot in the second building is typically referred to as the "backhaul."

802.11g

To keep up with the 54-Mbps speed claims of 802.11a, the 802.11g protocol was ratified in 2003. This protocol took the OFDM modulation technique of 802.11a and applied it to the 2.4 GHz spectrum of 802.11b. Because it operated in 2.4 GHz, it was possible to remain backwards-compatible with 802.11b equipment. 802.11g radios support both OFDM and DSSS modulation techniques. Therefore, an 802.11g device would, in theory, be compatible with an original 1 or 2 Mbps 802.11 DSSS device from 1997.

Keep in mind that a typical residential or small business hotspot has a DSL or similar connection behind it providing the bandwidth to the Access Point. These broadband connections typically provide speeds in the 1.5 to 3 Mbps range. Obviously, the bottleneck in a Wi-Fi deployment is usually the DSL (or even T1) pipe. Therefore, the advantages of higher speed wireless connections (such as 802.11g) are often limited because of the Internet connection. The only exception would be if there is a large number of data transfers between wireless clients and PCs on the local area network (or between two wireless PCs). In those cases (such as gaming or local file transfers), users will notice a significant speed increase when using 802.11g, compared to slower wireless protocols, such as 802.11b. In many large-scale community wireless networks, a system of repeaters will be used to enhance coverage in dead spots. Because each repeater (such as WDS) reduces the bandwidth by half, using 802.11g (and 54 Mbps) is often desirable. The logic here is that you can halve 54 Mbps more times then you can halve 11 Mbps, and yet still wind up with a useable, decent bandwidth speed for the client.

The pros and cons of 802.11g are as follows:

- Upside: Relatively fast speed; compatible with 802.11b
- Downside: Interference from other 2.4 GHz devices; only three non-overlapping channels

Ad-Hoc and Infrastructure Modes

When architecting an 802.11 network, there are two modes in which you can operate: Ad-Hoc and Infrastructure. In Ad-Hoc mode (see Figure 1.1), sometimes called "IBSS" or "Independent Basic Services Set", all devices operate in a peer-to-peer mode. There are no access points used in this topology, as all devices communicate directly with all other devices.

In Infrastructure mode, an AP is connected to a wired infrastructure (such as Ethernet) and all of the wireless devices communicate with the AP. Even if two wireless devices are located right next to each other, all communication between the devices occurs through an AP. When using Infrastructure mode, a collection of wireless devices connected to an AP is referred to as a Basic Service Set (BSS). If two or more BSSs are connected together using a "Distribution System" (such as wired Ethernet), the collection of BSSs is referred to as an Extended Service Set (ESS).

Figure 1.1 Ad-Hoc Mode

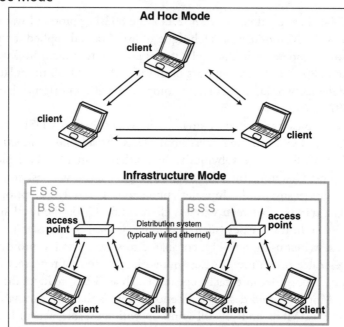

Connecting to an Access Point

In order to establish communication between a client device and an AP, the client must know the Service Set IDentifier (SSID) of the network and then perform two special steps: authentication and association. Therefore, the communication relationship exists in three states:

- Unauthenticated and unassociated
- Authenticated and unassociated
- Authenticated and associated

SSID Discovery

There are two possible ways for the client to know the SSID of the network. Either the SSID is told to the client by the AP (often called Open Network mode), or the SSID has to be known by the client via some other method, such as being preprogrammed into the client device by the system administrator (often called Closed Network mode).

The AP broadcasts "management beacons," typically every 100 milliseconds. These management beacons are a special kind of "Wi-Fi mating call," containing all of the synchronization information that the client needs to know in order to associate with the network, including channel, supported speeds, timestamps, WEP status, and other capability information. With Open Network mode (see Figure 1.2), the SSID is included in the management beacon.

Figure 1.2 Open Network Mode

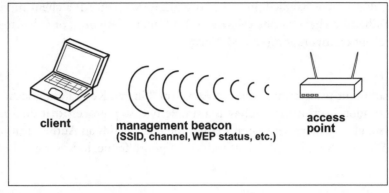

With Closed Network mode (see Figure 1.3), the client uses its preprogrammed knowledge of the SSID and broadcasts probe requests across all channels. The probe request includes the SSID of the network that the client is attempting to communicate with. If the AP hears a probe request on its channel, along with a matching SSID, then the AP will answer back with a probe response. This probe response will contain synchronization details, similar to the information found in a management beacon.

Note that APs in Closed Network mode still broadcast a management beacon; however, the portion of the management beacon where the SSID would normally be found is blank.

Figure 1.3 Closed Network Mode

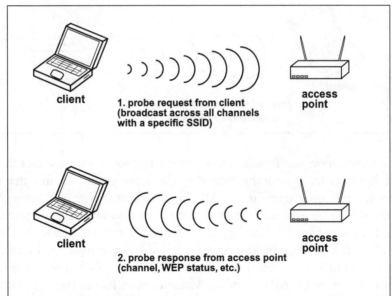

Again, the key distinction between these two methods is the way in which the Service Set IDentifier (SSID) discovery is handled. In order to associate with an AP, a client must know the SSID of the network. Either the client can be told the SSID (Open Network), or it has to be preprogrammed in the client configuration (Closed Network).

Authentication

Authentication can occur using either "Open System" or "Shared Key" authentication (see Figure 1.4). Null authentication, as its name implies, is a simple two-step process that does not require any credentials to be supplied. The process begins when the client sends an Authentication Request frame to the AP. The AP responds with an Authentication Response frame, indicating either success or failure.

Figure 1.4 Open System (Null) Authentication

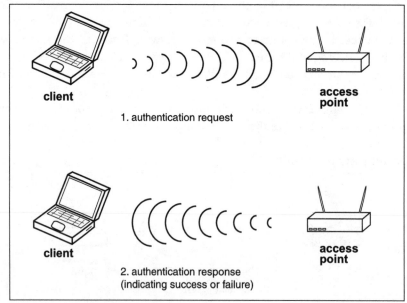

Shared Key authentication (see Figure 1.5) is a four-step process that involves the client's knowledge of the WEP key in order to be authenticated by the access point. The first step is that the client sends an Authentication Request frame to the AP. The AP responds with 128 bytes of challenge text. The client uses the WEP engine to encrypt the 128 bytes of random challenge text and then sends back a Challenge Response frame, containing 128 bytes of (encrypted) cipher text. In order to authenticate the client, the AP decrypts the cipher text and sees if it matches the original challenge text. This process is used to validate whether or not the client actually knows the shared secret of the WEP key. The final step is for the AP to send an Authentication Result frame, indicating success or failure.

Figure 1.5 Shared Key Authentication

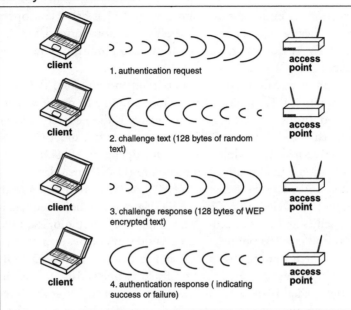

Association

Once the client has been authenticated, the next step is to associate with the access point (see Figure 1.6). The client sends an Association Request frame (including the SSID) and the AP sends back an Association Response frame, indicating success or failure.

Figure 1.6 Wireless Access Point Association

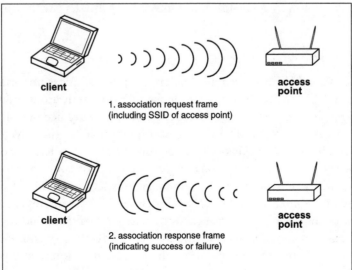

FCC Regulations

One of the reasons that so many enthusiasts are drawn to 802.11 is because it operates in an "unlicensed spectrum," where a license to operate from the FCC is not required. Unlike other wireless activities, such as radio or TV broadcasting, we do not need to purchase frequencies or pay a monthly service fee to use the airways for Wi-Fi.

While 802.11 operation is free from cost, it is important to remember that it is not free from regulations. The rules for operating RF equipment vary from country to country, depending on the local regulatory agency involved. In the United States, that agency is the Federal Communications Commission (FCC). The rules for operating 802.11b equipment fall under the 1985 Industrial, Scientific, and Medical (ISM) mandate, while the rules for operating 802.11a equipment fall under the 1997 Unlicensed – National Information Infrastructure (U-NII) mandate.

ISM regulations actually predate the existence of 802.11. ISM provides unlicensed spectrum in the 902–928MHz, 2.400–2.485GHz, and 5.725–5.850GHz ranges. These frequencies were put to work in a variety of capacities and introduced a large number of technological innovations such as pagers, microwave ovens, and cordless phones. The downside, of course, is that by the time 802.11 moved into the landscape, there were already a large number of users in the 2.4 GHz frequency space.

U-NII, on the other hand, was designed specifically for wireless data networking. If you think back to 1997, it was a time when there was a national movement to bring the Internet to schools across the country. Apple computer petitioned the FCC with the thought that the best way to "wire" the schools was to "unwire" them, and so the FCC granted frequencies in the 5.15–5.25 GHz, 5.25–5.35 GHz, and 5.725–5.825 GHz ranges specifically for this purpose. Smart decisions were made in an attempt to optimize the spectrum. For example, the "Lower Band" of four non-overlapping channels (5.15–5.25 GHz) was reserved for lower-power indoor devices (up to 50mW). The "Middle Band" of the four non-overlapping channels (5.25–5.35 GHz) was reserved for medium-power indoor/outdoor devices (up to 250mW). Finally, the "Upper Band" of the four non-overlapping channels (5.725–5.825 GHz) was reserved for higher-power outdoor devices (up to 1W). In this manner, long distance (high power) point-to-point links did not interfere with shorter range (lower power) wireless networks.

FCC and IEEE Regulations

The functions of the FCC and IEEE serve very different purposes and it is important to understand their distinctions. The FCC is the regulatory body for the telecommunications industry. Among other things, the FCC manages the airwaves by publishing regulations, issuing licenses, allocating the radio spectrum, and conducting investigations. The FCC is also responsible for the ISM and U-NII mandates. More information about FCC regulations can be found here: www.fcc.gov/oet/info/rules.

On the other hand, the IEEE is a professional organization whose mission is to create and develop standards (particularly networking and communications standards). The IEEE publishes these standards in order to promote interoperability between devices. Greater interoperability between vendors helps to create more choices for consumers and ultimately benefits the industry as a whole by encouraging greater usage and adoption rates by the public. The IEEE is responsible for the family of 802.11 protocol definitions. More information about the IEEE can be found at www.ieee.org.

When it comes to use of the airwaves in the United States, we find an area of intersection between the FCC and the IEEE. As you read the FCC regulations, you'll find some differences between the FCC rules and the IEEE specifications. This variance reflects the IEEE's position of remaining "conservative" and within the boundaries of the FCC mandates. For example, the 802.11a IEEE guidelines for power output limitations are actually 20 percent lower then the FCC limits. Table 1.3 shows the relative FCC and IEEE power output limits.

Table 1.3 802.11a Power Output Limits

Spectrum Range	FCC Limit	IEEE Limit
5.15 GHz–5.25 GHz	50 mW	40 mW
5.25 GHz–5.35 GHz	200 mW	250 mW
5.725 GHz–5.825 GHz	800 mW	1,000 mW

Why Wi-Fi?

So, you may be wondering: Why should I build a community wireless network? The answers to this question are as varied and diverse as the communities themselves that have deployed wireless networks.

Early models for building wireless networks focused around their commercial potential. For several years, a number of companies (both venture capital and privately funded) embarked upon the task of setting up as many for-pay hotspots as fast as possible across the country. This "land grab" mentality was met with limited success. While some market existed for paid Internet access in "captive audience" scenarios (such as hotels and airports), most venues are now discovering that the cost and hassle of the billing process make it more attractive to just offer Wi-Fi as a free service or "amenity" that compliments their business.

For example, coffee shops and hotels are not in the business of being an ISP. By offering free access, they can focus on what they do best—making lattes, yummy treats, and providing lodging for guests. Retail locations have now discovered that offering free Wi-Fi has enormous benefits as a marketplace differentiator. In other words, while the early models were to hope that Wi-Fi could be used as a revenue source, nowadays, these venues have made a significant change in perspective. They realized that millions of people had Wi-Fi built into their laptops, but a very small percentage were willing to pay for Wi-Fi when roaming outside of their home. With Wi-Fi revenue at miniscule levels, and deployment costs getting lower and lower (thanks to cheaper and cheaper hardware!), these venues understood that it made more sense to offer Wi-Fi for free as a way to attract more customers.

We've seen this paradigm before: hotels (like any industry) are always looking for a way to attract new customers and gain a competitive edge. Decades ago, hotels advertised these differentiators in bright neon lights. First is was "Air Conditioning," then "Color TV," then "HBO," and now "Free Wi-Fi."

Beyond the commercial applications, one of the most exciting areas of Wi-Fi proliferation has been in residential neighborhoods—in particular, in disadvantaged and low-income neighborhoods. However, regardless of your socio-economic status, the idea of sharing bandwidth and saving money has always been an attractive motivator. With the advent of 802.11, we can now legally share our Internet connection and get to know our neighbors better at the same time!

SoCalFreeNet

SoCalFreeNet.org is a non-profit community group dedicated to building, deploying, and growing public wireless networks in order to increase widespread broadband adoption and create an empowered, connected society, where technology becomes a community resource.

In San Diego, our model for SoCalFreeNet residential deployments works like this: The property owner pays for the hardware and monthly ISP service fees, while the user group volunteers provide the wireless expertise and physical deployment services.

Note...ISPs

Our policy is to only deploy community wireless networks using ISPs that allow bandwidth sharing. If the ISP prohibits sharing in their Terms of Service (ToS), we will not deploy a wireless network using that ISP. However, some ISPs allow (even encourage) bandwidth sharing. By using "Wi-Fi friendly" ISPs, we can ensure that our deployments are legal. At the time of this writing, companies such as Speakeasy.net and Megapath.net allow legal bandwidth sharing. A more complete list of ISPs can be found at www.eff.org/Infrastructure/Wireless_cellular_radio/wireless_friendly_isp_list.html.

Benefits for Property Owners

The advantages to deploying free wireless access are numerous. For a property owner, providing bandwidth is a way to "give back" to the community. Property owners can leverage their valuable rooftop locations for mounting antennas and other gear in order to provide a community resource for all to share.

In addition to the community benefit, property owners can also make their own properties (particularly rental locations) more attractive to potential tenants. Since a renter can avoid a monthly service fee for Internet access, the value of that particular property is greater then other locations where the renter would have to pay a monthly fee for bandwidth.

Deploying free wireless may (in certain circumstances) also be tax deductible for the property owner. Please consult with a tax professional for additional details.

Benefits for Volunteers

From a volunteer perspective, participating on a wireless deployment provides a way for members to "give something back" to the community. Most user groups have a community service focus and donating their time for community projects fits nicely into their missions.

Volunteering on a project also gives people the opportunity to get hands-on experience with wireless equipment. From a resume perspective, this kind of experience is tremendously valuable. Often times, the equipment used for these projects is very expensive. Therefore, volunteers have a chance to "play" with equipment that might otherwise be out of their reach. Finally, outdoor projects are fun! Going outdoors and working with wireless gear gives people a chance to meet likeminded wireless enthusiasts and work on their tans!

Social Ramifications

For most of us, we cannot imagine a daily existence that lacked basic necessities such as electricity, hot water, indoor plumbing, natural gas, and so on. However, people living 100 years ago would have considered these modern conveniences to be a luxury. Nowadays, we take these things for granted. Can you imagine living in a world without them? That sensation—a bizarre and barbaric "how could you live like that" feeling—is the way that people will view our lives 100 years from now. Bandwidth is quickly becoming a "fourth utility," comparable to electricity, water, and gas. In the future, people will just expect it to be there… anything less will simply be "uncivilized."

Many decades ago, computers were thought of as stand-alone tools. While first used as business devices, they eventually filtered into home life for utility and recreational purposes. As computing evolved, the idea that computers could interact and "network" together grew in popularity. Before the advent of the Internet, computers were networked to share files and devices, such as printers. Early networking attempts were focused on lowering hardware peripheral costs. For example, by networking an entire business you could deploy a single printer for a group of people to share, instead of having to install a separate local printer on each person's desk. Also, these people could now share files (such as word processing or spreadsheet documents) without having to copy them onto floppy disks and carry them from one location to the next. While this added some convenience, most applications remained "stand-alone" in their functionality.

With the advent of the Internet, computing has taken a radical shift in usage patterns. Modern-day computing power has shifted from the desktop to the network. Sure, a stand-alone PC can still do word processing and spreadsheets, but the network-centric paradigm of modern computing has increased the value of the network exponentially. Nowadays, when people walk up to a computer, they just expect to be able to open a browser and launch Google, check their e-mail, or get directions from MapQuest. Not being connected to the network is tantamount to the computer being "down" or inoperable. Even in an enterprise context, most modern business applications have a browser interface or some networked component. Centralized databases, information warehouses, and intranet applications have become the mainstay of any contemporary business.

From a community wireless networking perspective, delivering bandwidth to individuals is the true definition of "bridging the digital divide." Some years ago, we might have said that just putting a computer in every person's home was the key to advancing society. Now, clearly, a computer without an Internet connection is as good as a paperweight. Bringing bandwidth to the masses is the key to creating opportunities to learn, grow, and thrive. Wireless technology is just one of many ways to bring that bandwidth into homes and businesses around the world. It just so happens that wireless is also the easiest and most cost-effective transport method to accomplish the goal of delivering the bandwidth.

Security in a Community Wireless Network

One of the inherent features of wireless technology is that the RF signals don't stop at your walls. From an enterprise perspective, this is viewed as a weakness. However, from a community wireless networking perspective, this is viewed as our primary advantage and benefit. RF signals know no boundaries. We can't see them, but they're everywhere. This means that the wireless network that provides bandwidth to users in a community also has some inherent security risks that need to be considered. We'll touch on some of those issues here, as well as in Chapter 3.

NEED TO KNOW...SECURITY RECOMMENDATIONS

Most community wireless networks are intentionally unencrypted. Since all of your data is floating through the air, anybody within range to pick up the wireless signal can potentially read your data. 802.11 technology is analogous to using hubs in a wired network. All data is accessible by all hosts plugged into the hub (or associated with the wireless Access Point). Therefore, if your data is sent in cleartext, it may be compromised or monitored. The simple solution here is to use higher-level encryption. Here are some suggestions:

E-Mail: www.fastmail.fm

Web Surfing: www.freedom.net, www.anonymizer.com, any https Web site

File Transfer: scp (instead of ftp) www.chiark.greenend.org.uk/~sgtatham/putty/

Remote Shell: ssh (instead of telnet) www.chiark.greenend.org.uk/~sgtatham/putty/

Every Computer Needs to Be Protected

Firewalls placed between the DSL/Cable connection and the community wireless network can be configured to block typical attacks coming from the "outside world." For example, the wireless network can be protected from port scanning, worm attacks, and other malicious activity coming from the Internet by enabling a firewall at the point of entry of the DSL/Cable connection.

However, computers inside the wireless cloud are all still visible to each other. It's as if they are all plugged into the same hub and operating on the same network. If the computer has an IP address, it is "visible" to the other computers in the same wireless network. Therefore, each and every computer needs to protect itself with a host-based firewall. Windows XP and all Linux/Unix flavors have this functionality built in. However, other operating systems can add it using third-party applications such as Zone Alarm or Norton (Symantec) Personal Firewall.

Even if you "trust" all of your neighbors, you simply never know when an attacker will come driving through your neighborhood and will be unable to resist the temptation to sniff the traffic and start probing visible machines on the network. Installing a personal firewall will limit your risk of exposure to these kinds of attacks. Most consumer-grade APs include firewall functionality; however, it is important to note that this firewall exists between the WAN port and all the LAN/Wireless devices. In other words, most APs treat the LAN (typically a four-port switch) as if it is on the same subnet as the other computers connected wirelessly. The firewall does not protect the LAN computers from attacks generated on the wireless segment, nor does it protect wireless devices from attacks generated by other wireless devices. In most cases, the AP simply considers the LAN and wireless segments to be "trusted." This is another reason why each computer on your network should have its own host-based firewall: to protect itself from other unauthorized devices.

In addition to installing a firewall, it is always a good security practice to make sure your computer's system patches are up-to-date. Further, you should utilize anti-virus and anti-spyware applications and always update the definition files for those applications. These steps will help to protect your system against virus and worm attacks.

Wireless users need to be concerned about any user who is in range of their Access Points. However, the common misconception is that this threat is limited to nefarious individuals lurking in the parking lot. In reality, the threat is much greater, as wireless signals could potentially be intercepted (or injected) from miles away. With line of sight and the right equipment (a high-gain directional antenna and an amplifier), it is possible for an attacker to pick up wireless signals 20 to 25 miles away.

Legal Liability

One of the unfortunate downsides to any open wireless Access Point is the potential for it to be abused for illegal and immoral activities. Community wireless networks need to be concerned about activities such as hacking attacks, virus/worm launching, SPAM, e-mail fraud, and illegal downloads (this includes child pornography, copyrighted materials like music and movies, and so on). Anytime you consider deploying an open AP, there are both legal and moral issues that need to be considered and addressed.

Most community wireless networks use Network Address Translation (NAT) as their gateway between the wireless network and the wired backbone. NAT'ing is used to share the single IP address typically provided by the DSL or cable company. During an investigation, law enforcement will typically obtain logs from the victim's computer and attempt to trace the activity back to the suspect using the IP address as a starting point. By serving the ISP with a search warrant, the name and address of the individual owner of the Internet account can be obtained. Because of NAT'ing, all of the traffic from the wireless network appears to come from a single IP address, thus providing the cloak of anonymity to the perpetrator. Unfortunately, the illegal traffic appears to come from the IP address of the DSL/Cable modem. Therefore, the innocent owner of the AP becomes the unknowing suspect of an investigation.

A variety of investigative techniques are used by law enforcement to avoid kicking in the door of the wrong "suspect," who is, in actuality, really a victim themselves. On the other hand, law enforcement is concerned about criminals who claim to be a victim simply because they are running an open Access Point. Good computer forensic work can usually provide evidence and help determine additional facts in a particular case. Serving a search warrant and arresting the wrong person is a nightmare scenario for law enforcement since it creates unnecessary liability for the investigative agency and also puts agents in harm's way during the search. For example, what would happen if a search warrant was served and it resulted in a physical altercation or unintentional discharge of side arms? Accidents can happen and safety is always a concern for both citizens and members of law enforcement.

Defending the Neighborhood

Builders of community wireless networks are motivated by creating community resources and sharing bandwidth in safe and legal ways. As such, these builders (almost always volunteers) have no interest in seeing their hard work being used as a safe harbor for criminal activities. Nobody wants the network used as a tool for illegal downloads or hacking activities. Therefore, it is very important to establish and maintain good relationships with the law enforcement community.

Building bridges with law enforcement agencies helps them to understand the mission of community wireless networks, and helps us to understand the needs of law enforcement during an investigation. If illegal activity occurs on a community wireless network, law enforcement should not need to kick in any doors. Rather, a simple phone call to the designated contact should yield a willing partner to assist in an investigation. To facilitate this kind of community partnership, we recommend the establishment of a "Wi-Fi Neighborhood Watch" program. Following the model of the traditional neighborhood watch program, established to protect the neighborhood from burglary and violent crime, the mission of a Wi-Fi Neighborhood Watch should be to keep the Internet safe and serve as a powerful message that your neighborhood is not a place to perpetrate Internet crimes. Community wireless networks are part of the public domain. As such, there is no expectation of privacy (no different then a community park or a public sidewalk).

To protect your network, there are a number of steps you can take. The first is to establish a captive portal. A captive portal is a method whereby, when a client opens a Web browser, the captive portal directs them to a specific Web page, regardless of what Web page is initially requested by the browser. So, when the user opens a browser, instead of going to their start page, it automatically redirects them to a page where the network's Terms of Service (ToS) are displayed. Typically, at the bottom of the page is an **OK** or **I Agree** button, which the user must click before they can continue. Unless they agree to the terms of service, the user cannot gain access to any Internet resources. Once they click **OK**, they have acknowledged the ToS and access is granted. Chapter 3 will include more specific instructions for setting up a captive portal along with other security information.

Summary

In this chapter, we reviewed the history and basics of 802.11, including the differences between 802.11a, 802.11b, and 802.11g variants. 802.11g (54 Mbps) is backwards-compatible with 802.11b (11 Mbps). They both operate in the 2.4 GHz spectrum, with three non-overlapping channels. 802.11a (54 Mbps) operates in the 5 GHz spectrum, with 12 non-overlapping channels. Speed ratings for 802.11 represent theoretical signaling rates and not actual data throughput, which is typically half of the advertised speeds.

The FCC regulates the airwaves in the USA. The ISM (1985) bands of frequencies includes 2.4 GHz and is used for 802.11b/g. The U-NII (1997) bands of frequencies includes 5 GHz and is used for 802.11a. Since the 2.4 GHz ISM band is in use by a variety of devices such as cordless phones, microwave ovens, and wireless video cameras, these frequencies are often referred to as the "junk band." By comparison, 5 GHz is less crowded.

Building community wireless networks has numerous benefits for property owners, as well as volunteers. The SoCalFreeNet project is dedicated to building community wireless networks and using technology as a tool to enhance communities. Bringing bandwidth to neighborhoods (particularly disadvantaged neighborhoods) can greatly help bridge the digital divide.

From a security perspective, it is important to remember that all wireless packets float through the air and, hence, are vulnerable to potential eavesdropping and injection attacks. Most community wireless networks are not encrypted, therefore extra attention must be given to educating users about the security risks and ensuring that all devices incorporate a host-based firewall. Use of a captive portal with terms of service is also a necessity to limit the property owner's liability. Finally, you must be sure that bandwidth sharing is not a violation of the terms of service of your ISP.

## Chapter 2

SoCalFreeNet.org: Building Large Scale Community Wireless Networks

Topics in this Chapter:

- Wireless Distribution System (WDS)

- 5-GHz Links

- Working with Client Devices

- Competing with the Phone/Cable Companies

- Outfitting Coffee Shops and Retail Locations

- Getting the Neighborhood Involved

Introduction

Setting up a community wireless network can happen in a variety of topologies. Some networks are small and simple. Others are large and complex. For example, you could take your home cable/DSL connection (as long as sharing is allowed by your ISP!), connect it to any Access Point, and—voilà!—an open AP! By not using WEP (or some other encryption), you are leaving your Access Point open for sharing. Anybody in range of your wireless signal can associate with your network and utilize your bandwidth. While you can expand your range using higher-power radios and external high-gain antennas, there are some limits to the coverage you can achieve with a single AP.

Alternatively, if you want to cover an area that extends beyond the range of a single AP, there are a number of solutions available at your disposal. You will have to consider two issues. The first is where to deploy APs to get coverage to all of the clients. The second is how to get bandwidth to the APs.

The APs that provide 802.11 coverage to the clients (typically 802.11g or 802.11b) are often called "client access radios." Ideally, you should place APs close enough to each other that you still have a little bit of overlap (usually 10 to 15 percent) to provide seamless coverage. Although the same Service Set ID (SSID) is used by each AP, you should always put adjacent APs on different channels to avoid interference. Of course, Wireless Distribution System (WDS) breaks this rule. More on WDS in a moment.

Next, you have to consider the issue of how to deliver bandwidth to each AP. (This link is often called the "backhaul.") The backhaul can occur wirelessly (such as with WDS), or it can be achieved via a 5-GHz link. Alternatively, you can learn more about Mesh Networks in Chapter 9. Using a second 802.11 b/g channel as a backhaul is possible, but not recommended since we have found long-distance links in 2.4 GHz to be unreliable in major metropolitan areas, due to interference from other devices.

In this chapter, you will learn about ways to expand the traditional concept of Wi-Fi as a wireless local area network (WLAN) into a wireless wide area network (WWAN). Using the technology in ways not intended by the original protocol often pushes the edge of the envelope in terms of performance and usability. However, as a wireless hacker, you are probably eager to roll up your sleeves and start experimenting with ways to grow your outdoor footprint. Also in this chapter, we will talk about some of the social and community elements of wireless networking, as experienced by SoCalFreeNet. SoCalFreeNet is a non-profit group dedicated to building public wireless networks. I highly recommend joining (or starting) a wireless users group in your area and working together to build community wireless networks.

Wireless Distribution System (WDS)

In the 802.11 literature, an Access Point that bridges between wireless and wired segments is connected to other APs using what is referred to as a "distribution system". This distribution system is typically a standard cat5 Ethernet connection.

In the previous chapter, we described an AP with associated clients as a "Basic Service Set" (BSS), and we described two or more BSSs connected together as an "Extended Service Set" (ESS). Instead of connecting the APs in an ESS together using wired Ethernet, the idea of connecting multiple APs together via a wireless link has gained a great deal of attention lately because the deployment model is so much easier. In this manner, we replace the Ethernet "Distribution System" with a Wireless Distribution System. Hence, the term WDS.

APs that support WDS can communicate with other APs wirelessly, while still communicating with client devices at the same time. Not all APs support WDS; however, this feature is becoming more and more popular among wireless devices as users discover the benefits of the simplified deployment model. With WDS, you don't need to run Ethernet cables to each of your APs.

The downside to using WDS is that bandwidth is cut in half every time a frame travels through a WDS AP (sometimes called a "hop"). This happens because all WDS devices must operate on the same channel. Therefore, the WDS AP must listen for the frame, and then transmit it on the same channel.

Another problem with WDS is that it is not universally supported by all Access Points. Furthermore, some WDS APs are not compatible with other WDS APs. WDS compatibility continues to improve, but to be sure your WDS devices are interoperable, you should do some Google searching, or perform independent testing on your own before you commit to a specific vendor solution.

WDS seems to be a good solution to extend a wireless network a hop or two in a small residential environment. For example, if you have a particularly large home that doesn't seem to get adequate coverage from a single AP, you might consider extending your wireless coverage using WDS by placing one AP upstairs and the other AP downstairs.

5-GHz Links

One of the challenges to deploying a network in any major metropolitan area is the level of Radio Frequency (RF) congestion at 2.4 GHz. Some of our earliest SoCalFreeNet deployments utilized two 802.11b radios at each rooftop node. One radio would provide client access to the surrounding neighborhood with a high-gain omni-directional antenna. The other 802.11b radio (on a different channel) would utilize a directional antenna pointed back to our main rooftop "node" (where the DSL terminates). Our initial thought was that separating the client access and backhaul radios onto different channels would improve our performance. At first, this configuration proved to be very effective. However, over time, the noise floor at 2.4 GHz increased to the level where the backhaul links became unreliable. This happened because of the large number of residents who deployed their own Access Points. Noise from other sources also contributed, such as cordless phones, wireless cameras, and other devices operating at 2.4 GHz.

To solve the congestion problem, we decided to migrate our backhaul links to 802.11a in the 5-GHz spectrum. Initially, these links operated at 5.3 GHz using out-of-production Proxim 8571 802.11a APs. These devices could be purchased for as little as $20 from aftermarket resellers. On the other side of these links, we used Soekris devices running Pebble software as an 802.11a client (using a second 802.11a PCMCIA card which we "harvested" from another Proxim 8571). Later, we upgraded many of these links using Soekris devices containing mini-PCI 802.11a radios, thereby allowing us to upgrade these links to 5.8 GHz. By moving to 5.8 GHz, we were able to use higher-gain antennas and remain FCC compliant.

Working with Client Devices

In our earliest SoCalFreeNet deployments, we would help residents get online using any device available. Typically, this meant PCMCIA cards, USB client devices, or whatever the resident wanted us to install on their computer. The problem with this method was that the moment we touched a resident's computer, we seemed to "own" that device. Inevitably, a problem would arise and even if that problem had nothing to do with wireless, we would always get phone calls and e-mails asking for help. From an end-user perspective, I can understand why this happens. One day a bunch of wireless geeks come over to your house and install something new. The next day your printer stops working, so naturally it must have been something caused by the new wireless card.

It would be great if we had the ability to service all the non-wireless technical support issues in the neighborhoods we operate in. Unfortunately, given our limited time and resources, we found it necessary to restrict our support services to maintaining the wireless infrastructure and leaving the end-user technical support issues to third-party companies or relying upon "neighbor helping neighbor" methods.

In order to reduce the number of non-wireless technical support issues, we decided to take a "hands-off" policy with respect to client PCs. In other words, we try not to touch a resident's computer if possible. To accomplish this task, we set about creating a standardized client kit for connecting to a SoCalFreeNet network. Our chosen design was to use a D-Link DWL-810+ Ethernet-to-Wireless Bridge, with a D-Link DWL-R60AT antenna.

The D-Link DWL-810+ device acts as an 802.11b client and then has an RJ-45 standard Ethernet port as its "output." These devices are typically used to bring network access to set-top boxes (e.g., Tivo), gaming consoles (such as X-Box), or printers that are expecting a wired Ethernet interface. The nice part of using an Ethernet bridge is that the neighborhood PCs almost always have some kind of wired Ethernet jack, which is typically configured (by default) to use Dynamic Host Configuration Protocol (DHCP).

Since it's Ethernet, you can place the D-Link bridge in a window and run up to 100 meters of Cat5 cable to the computer. Unlike the USB clients we had used in the past (and their 15-foot USB cable limitations), using Cat5 gave us a great deal of deployment flexibility. We would simply tell people to put this device in a window sill, point the antenna at one of our rooftop nodes, and then run a cable to their PC.

Another nice feature of the D-Link DWL-810+ is the ability to support external antennas. Since the D-Link has an RP-SMA connector, we had some different options for detachable antennas. The DWL-R60AT provides 6 dBi of gain in a nice, small form factor that fit nicely with the DWL-810+.

NEED TO KNOW...D-LINK PRODUCTS

For more information on the DWL-810+, visit www.dlink.com/products/?pid=21.

For more information on the DWL-R60AT, visit www.d-link.com/products/?pid=57.

Note that D-Link also supports 802.11g using the DWL-G810 (www.dlink.com/products/?pid=241).

In short, you plug in the kit and it works, with no user intervention. Unlike our earliest deployments, which required a volunteer to go to somebody's home and hand-hold a resident through the wireless installation process, we found that the kits practically installed themselves. We include a simple instruction sheet and found that residents were pretty self-sufficient. Additional help is rarely needed. It's a perfect "zero truck roll" solution for our group. Figure 2.1 shows a picture of the D-Link "kit." Figures 2.2 and 2.3 show an enhanced 2.4 GHz Ethernet to wireless bridge and an indoor 6dBi Microstrip Antenna. (Images courtesy of D-Link)

Figure 2.1 SoCalFreeNet D-Link Kit

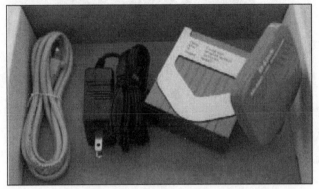

Figure 2.2 Enhanced 2.4 GHz Ethernet to Wireless Bridge

Figure 2.3 Indoor 6dBi Microstrip Antenna

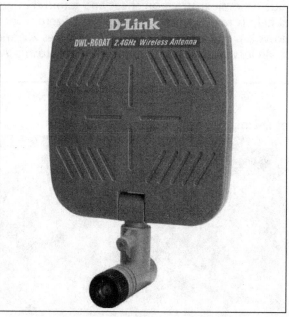

Competing with the Phone/Cable Companies

I always like to remind people that SoCalFreeNet is not competing with the local phone/cable companies or other bandwidth providers. The reason is because we offer bandwidth for free and provide two completely different service models.

SoCalFreeNet offers a "best effort" service with no guarantees of reliability or "Service Level Agreements." The commercial providers cost money, but in exchange, they provide around the clock monitoring, and a telephone hotline so you can yell at somebody when your connection is down.

Although we spend an enormous amount of time and energy building redundancies and reliability into our network, the "best effort" model means that nobody carries a beeper or is on call 24/7 to attend to the network. We do the best we can, given the resources and volunteers that we have.

Sure it would be nice to have a 24/7 Network Operations Center and we acknowledge that we aren't operating a "perfect network," but we'd like to think that "best effort" is still better than "no network" and in many of our neighborhoods (particularly the low income and disadvantaged ones), "no network" is often the only option.

While rare and infrequent, network problems do occur. However, one of the nice surprises we discovered is that being "free" does have its advantages. People know that you get what you pay for and therefore, their level of expectation is generally set pretty low. As such, given the free nature of SoCalFreeNet, people tend to be more accommodating of unscheduled network outages.

Outfitting Coffee Shops and Retail Locations

While much work is being done in residential neighborhoods, SoCalFreeNet also spends considerable effort to bring free wireless access to coffee shops and other retail properties.

On a recent car ride, while searching for a good lunch spot, I happened upon a local pizza shop with a sign out front reading "Free Wi-Fi for Customers." Curious, I walked into the store and had an enlightening chat with the business owner about their wireless deployment. It turns out that the owner of the business noticed that the local Starbucks store (two doors down) was offering T-Mobile hotspot service for a fee, so they decided to offer it for free as a service to customers and to attract more business.

After assuring me that he had obtained permission from the DSL company, (many business-grade DSL services do allow sharing), he proceeded to show me the Linksys box sitting in the corner, servicing the Wi-Fi needs of his patrons. To accommodate the "for customers only" aspect of the offering, the owner was using WEP and writing down the key on a piece of paper, which he manually handed out to each customer who inquired. Needless to say, this method of authentication is cumbersome at best.

I've also seen coffee shops attempt to perform access control by using "Closed Mode" (turning off the SSID broadcasts) as a method of authentication and manually distributing the SSID on a piece of paper to customers who desired Wi-Fi access.

While the "free with purchase" concept does have its merits, using "Closed Mode" or WEP to limit access to the network are poor choices and cause more problems than they solve. The primary shortcoming is that it involves user intervention and adds a layer of complexity to the user process. Customers should be able to open their laptop, launch a browser, and enjoy the Internet as quickly and easily as possible.

When I asked these business owners about support issues, I discovered that there was always a large percentage of customers who had no idea how to configure WEP, or enter a closed-mode SSID into their computer. This created a distraction for the business and an aggravation for the customer. In addition to the customer support problems, both of these methods require maintenance by the business owner, who has to change the WEP key or modify the closed mode SSID on a periodic basis. Rather than playing IT and technical support roles, most business owners would prefer to focus on running their business, making pizzas and blending frothy iced drinks.

In addition, creating an open Access Point creates a liability for these business owners in the event their network is used for illegal activities. Most property owners are not even aware of the risks involved. Community Wireless groups can assist these companies (oftentimes small coffee shops or local restaurants), by educating them about these issues and also helping them in hardware selection and with network deployment. We can help install captive portal systems and provide guidance on the proper use of Terms of Service agreements to help limit liability issues.

Another feature of particular interest to these kinds of installations is the ability to provide access "free with purchase" and offering the service free to customers who patronize the establishment. This also helps ease their concerns about laptop users who "camp out" and take up space for long periods of time. To solve these problems, there are solutions available, such as the "Pre-Paid Module" from

Sputnik. With this module, property owners can create one-time-use passwords (which they can print out on business cards) that allow access for any predetermined amount of time. See Chapter 8 for more details on this exciting product. While this kind of functionality is not necessary for residential-based networks, many business owners find this feature incredibly useful.

Getting the Neighborhood Involved

Whether working on a commercial or a residential deployment, the key to success is finding ways to get your community involved and letting them "take ownership of the network." On all of our deployments, we always invite members of the community to take part in the installations. After all, it's their network, so we actively encourage residents to participate. In addition, we typically identify a "network nanny" or "community contact" for each neighborhood we operate in. This individual helps to get residents online, and can also be extremely helpful in troubleshooting scenarios where physical access to the equipment is necessary. Finding the net nanny is pretty easy. It's usually that person who shows up on the first neighborhood install day, with a laptop in one hand, a Wi-Fi card in the other, and a big smile on their face.

Getting the word out about the network usually takes place on its own by word of mouth. However, handing out fliers can help increase network adoption. In each neighborhood, we typically recommend an "installfest" or "help the neighbors get online" day, at a predetermined time and place each month (for example, every second Saturday of the month from 9A.M. to 11A.M.). This can take place at a local coffee shop, or any familiar landmark. In addition to helping residents get online, it's also a great opportunity for volunteers to network and "geek speak" on the latest wireless industry happenings.

In one of our neighborhoods, we've even entered into a partnership with a popular coffee shop, called Influx Café. We provide D-Link kits, preprogrammed for use on our network. For residents who just can't wait for an installfest, they can head over to Influx and pick up a Wi-Fi kit and install it on their own. We sell the kit at cost through the coffee shop. Residents pay for the kit, and Influx remits the cash back to the group. If residents have any problems installing the kit, they can come back during one of our monthly "help the neighbors get online" days and a volunteer can provide personal assistance. Residents can also return their kit for any reason (within 30 days) at one of these monthly get-togethers.

Note...It Takes a Village to Route a Packet

A community wireless network, like any community resource, needs to be cared for by the neighborhood and takes time to develop. When a community embraces a cause, a certain kind of magic takes hold. It's no different than when a group gets together to clean up an abandoned park, or organizes crews to paint over graffiti on the streets. We've found that each neighborhood is unique, but no matter where we go, we always seem to find people who take pride in their community and embrace a project that helps improve the area and make it a better place to live.

The special bonus of Wi-Fi is that it is, by definition, a communication tool. Building community wireless networks is a process of building communities. Unlike typical computing activities which seem to "isolate" people, it turns out that in the process of building out the networks, we've created more human "touch points" between neighbors than we ever imagined. This is not just "digitally," via e-mail and Web exchanges that are now possible, but also in the "analog" (real) world, as we move "beyond" the computer screen, involving more and more people in deployment activities, helping the network grow, and encouraging neighbors to assist each other with wireless support issues. As a result, Wi-Fi has weaved its way into the fabric of the community.

Summary

In this chapter, we reviewed WDS, showed how to use 5 GHz as a backhaul option, and explained D-Link "kits," used for client access. We also talked about helping coffee shops/retail locations set up free hotspots. The wireless hacks described throughout this book represent the many lessons learned by SoCalFreeNet in our efforts to build public wireless networks. Knowledge sharing is an essential element in cultivating your volunteer base and expanding your network coverage. We are here to help one another expand our technical skills, share new ideas, and have fun!

## Chapter 3

Securing Our Wireless Community

Topics in this Chapter:

- The Captive Portal
- Building a PPTP VPN
- Hacking the Mind of a Wireless User
- Others Hacks

Introduction

At SoCalFreeNet, we design our networks to serve as a place for people in the community to relax and enjoy a stroll across the Web. Browsing pages securely across our wireless network helps create an atmosphere of community, much like a park. The park metaphor is often used by our group to embody the appropriate goals when building and maintaining our networks.

We designed our community wireless network to allow ordinary people to connect using basic wireless equipment. This emphasis on sticking with the basics enables access for broad range of users, from the technical elite to a grandparent experiencing the Internet for the first time. New users of wireless technology can only handle a certain amount of learning before the technology overwhelms them, so we strive to keep these users engaged within set boundaries on the level of security complexity.

However, wireless security clearly comes with issues. Problems with wireless security stem largely from the broadcast nature of the wireless signals. Computer signals sent across traditional wired networks flow in a far more controlled manner as switches, routers, and firewalls act as gatekeepers watching over the network. Our wireless networks, on the other hand, blast the packets out in all directions; the packets only stop when the signal finally degrades enough to prevent the equipment from *hearing* our signal. With high-powered receivers, some users can listen to wireless networks from miles away (one member set up a desert link over 13.5 miles long using Linksys equipment without any amplification). This also means some users may try to eavesdrop on our network from great distances.

These long distance eavesdroppers remind us of an age-old reality, that some users use (wireless) technology to achieve positive goals, while others use our network for negative ends. These nefarious users can often create problems for other individuals as well as the architects of our wireless community. And, like a park, we must take steps to ensure the community stays clean—thus, we pick up the trash.

These malicious users create problems in three main ways:

- Attacking other users in the community

- Attacking users or computers out on the broader Internet

- Committing crimes on our community network

These crimes manifest themselves when a user tries breaking into/hijacking computers, or attacks other users on the Internet. Alternatively, they may use the Internet connection to download child pornography or conduct illegal business.

All these activities potentially create havoc for the community, and may even create liability issues for the wireless architects. So our architects went to work finding solutions to these problems. While security is always the responsibility of the end user, there are some optional steps that can be taken to improve the safety of the user experience. In this chapter, the following topics for improving security will be discussed:

- Enabling Our Captive Portal

- Writing Our Terms of Service (ToS)

- Captive Portal Graphics

- Building a PPTP VPN

- Enabling Our VPN

- Configuring Our Community Users

- Hacking the Wireless User's Mind

- The Beginning and the End

Learning how to protect our community forms the heart and soul of this chapter. The material focuses on innovative ways to approach the security issues we face in our wireless park community.

The Captive Portal

The threat of litigation presents one of the largest barriers to creating secure community networks. Along with litigation, networks present a number of security challenges. While the security described here addresses many issues and mitigates a range of risk, these measures only work when the community takes advantage of them. Many users simply want the fastest path to connectivity and will blindly ignore security. This ignorance inspires the adage, "You can lead a horse to water, but you can't make it drink."

The captive portal concept flows from the legal issues surrounding wireless security. Beyond security, the portal offers a convenient way of providing information and guidelines to members of the community. This sharing tool works by sitting in the middle of users' Internet connections and intercepting their first connection to the Web. When the user's first connection request comes through, the captive portal redirects the user to a page containing the community's legal disclaimer and guidelines. This page, sometimes referred to as Terms of Service (ToS), forms the core of the community network's legal agreement with users. Here the user may read about the security implications of the network. SoCalFreeNet leverages this page to encourage users to take preventative security measures and use some of the advanced security features we provide.

NEED TO KNOW...LEGAL LIABILITY

Providing free wireless comes with some legal implications. Protecting the community from litigation represents an important feature of captive portal technology. The captive portal also serves as an important educational tool for users to learn about security and their own responsibilities in this area.

Preparing for the Hack

Many of the features described in this chapter rely on an open-source firewall solution called m0n0wall (m0n0wall is further discussed in Chapter 6). The m0n0wall team consists of a community of developers who came together and built a new style of Unix that uses a nice Web-based configuration engine. This graphical interface makes for easy configuration of m0n0wall's captive portal and

Point-to-Point Tunneling Protocol (PPTP) Virtual Private Network (VPN). The PPTP-VPN acts as an encrypted tunnel that users may employ to protect themselves from other users eavesdropping on the network. We will demonstrate how to leverage both features to enhance security for our community wireless network.

Preparing our network for a captive portal depends on the following advanced steps:

1. Wiring the network for security

2. Choosing the captive portal software and hardware

3. Installing and configuring m0n0wall

Wiring the Network for Security

Building a secure wireless network starts with a secure architecture for our traditional wired network. The network preference of limiting access to only the necessary services acts as the guiding principle, better known in the security world as "default deny" or "least privilege." Attaining these goals starts with a good overall design. Figure 3.1 portrays the network layout for SoCalFreeNet's node0 community network, an example of one possible way to set up a simple, single-node residential hotspot.

Figure 3.1 An Example of a Secure Network Architecture

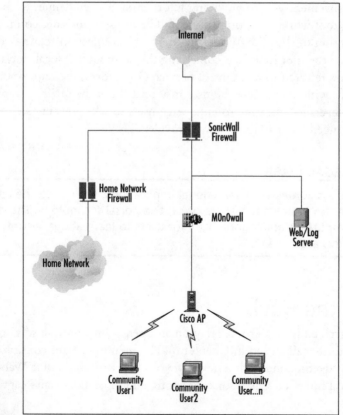

This network design protects the community members who provide the community network. The m0n0wall server acts as a captive portal and PPTP VPN concentrator for the wireless clients. The Cisco Access Point (AP) serves as a basic client access radio. After buying this particular Cisco 1120, we realized the device did not support an external antenna. So, we promptly modified the AP with an external SMC adapter. The modified Cisco is connected to a 15.3 dBi omni-directional antenna on the roof of the member's house. The house and antenna are missing from the diagram, but this shows the logical infrastructure supporting this node of SoCalFreeNet. Note that some SoCalFreeNet nodes are single independent APs in a resident's home, while other nodes are part of large-scale "hub and spoke" relay systems (as described in Chapter 2). APs can also be deployed as part of the Sputnik solution (as described in Chapter 8).

WARNING: HARDWARE HARM

Make sure to properly weatherproof all external connections. Water may seep into connectors and cause signal loss over time. There are many different opinions on proper weatherproofing techniques. The following link gives one possible approach: www.qsl.net/n9zia/wireless/sealing_andrews_connectors.html

The network diagram in Figure 3.1 only depicts the connections at a basic level. A complex set of rules exists to ensure that community users can only traverse out to the Internet, while still allowing administrators to manage the infrastructure. Protecting the environment requires proper configuration at all layers.

NEED TO KNOW...VOIDING YOUR WARRANTY

Modifying your hardware to add an external antenna adapter voids your Cisco warranty. This particular AP met our requirements for power output, but lacked the external antenna capabilities... in all honesty, we just wanted to see if we could do it.

Choosing the Captive Portal Software and Hardware

The next logical preparation involves selecting the hardware and software to serve as the captive portal. The captive portal serves the primary function of delivering ToS information when a user first attempts to connect. Delivering this information to the user may take many different technical forms.

Since many of our wireless projects face power constraints, we are limited to captive portals that support embedded devices. Embedded devices act as miniature computers hosting a variety of lightweight versions of UNIX. Many of these versions support the Soekris hardware devices (www.soekris.com). Here is a sample of some of the hardware on these nice little devices:

- **net4501-30** 133 MHz CPU, 64MB SDRAM, 3 Ethernet, 2 Serial, CF socket, 1 Mini-PCI socket, 3.3V PCI connector

- **net4521-30** 133 MHz CPU, 64MB SDRAM, 2 Ethernet, 1 Serial, CF socket, 1 Mini-PCI socket, Dual PC-Card socket, PoE

- **net4801-50** 266 MHz CPU, 128MB SDRAM, 3 Ethernet, 2 serial, USB connector, CF socket, 44 pins IDE connector, 1 Mini-PCI socket, 3.3V PCI connector

As you can see, these machines come with many of the features of a small computer packed into a tight package. These neat devices allow us to provide advanced features while keeping size and power requirements to a minimum. For this particular implementation, we chose a net4801. The extra horsepower provided by this model helps address the overhead for our PPTP-VPN discussed later in the chapter. Figures 3.2 and 3.3 show the external and internal view. More details on Soekris hardware can be found in Chapter 4.

Figure 3.2 External View

Figure 3.3 Internal View

Once we finalize our hardware selection, the software choices follow quickly. With the selection of an embedded device, our focus narrows to Pebble (also covered further in Chapter 6) and m0n0wall. Both of these distributions support captive portals. Pebble includes NoCat, while m0n0wall wrote their own. The Pebble software is a nice Debian distribution including HostAP drivers, a Dynamic Host Configuration Protocol (DHCP) server, a DNS server, Web server, and, of course,

SSH. m0n0wall followed the firewall mold and chose to support routing functionality, NAT, DHCP, IPSec/PPTP, DNS caching, DynDNS, SNMP, wireless drivers, and traffic shaping.

- **NAT** Network Address Translation. Mechanism for reducing the need for globally unique IP addresses. NAT allows an organization with addresses that are not globally unique to connect to the Internet by translating those addresses into the globally routable address space. Also known as *Network Address Translator*.

- **DHCP** Dynamic Host Configuration Protocol. Provides a mechanism for allocating IP addresses dynamically so that addresses can be reused when hosts no longer need them.

- **DNS caching** Provides a mechanism to hold DNS information in memory and speed up repetitive DNS queries.

- **DynDNS** Allows the device to use an external DHCP address while still offering services that require static IP addresses (Web server, mail server, etc.). This is generally done in conjunction with a Dynamic DNS vendor to achieve seamless service for DHCP users.

- **SNMP** Simple Network Management Protocol. Network management protocol used almost exclusively in TCP/IP networks. SNMP provides a means to monitor and control network devices, and to manage configurations, performance, and security. SNMP is also great for collecting statistics (more in Chapter 7).

Another network option for embedded systems is the Linux Embedded Appliance Firewall (LEAF) at http://leaf.sourceforge.net. This distribution provides network layer services and can support the NoCat captive portal. However, the captive portal must be installed separately. The Pebble images come with NoCat bundled and pre-installed.

While the Pebble solution offers some great flexibility and is very useful in multinode scenarios (particularly because of its atheros chipset support), in smaller scale scenarios we often choose m0n0wall because of its simplicity and excellent user interface. m0n0wall packs a lot of power into an 8-Megabyte (MB) Compact Flash (CF) card, and comes with a nice management interface, as seen in Figure 3.4.

Figure 3.4 The m0n0wall Management Interface

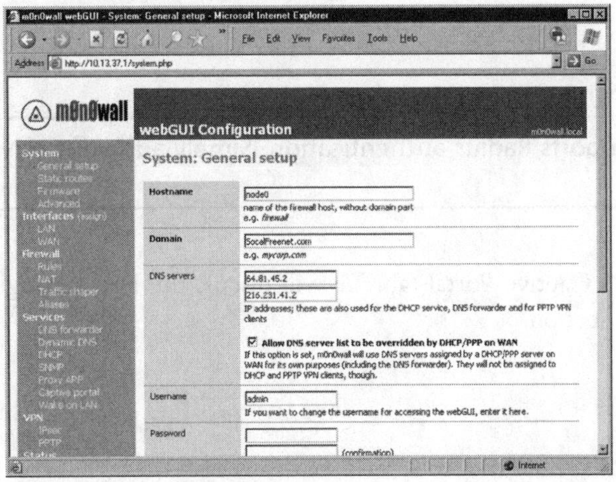

With the hardware and software selection finalized on m0n0wall, we simply install the firewall. The m0n0wall site (www.m0n0.ch/wall/installation.php) contains excellent instructions on this relatively easy process. Basically, you burn the Soekris image to a CF card, plug in the CF, and power up. The Soekris Web-based interface defaults to 192.168.1.1/24, with DHCP turned on. To begin managing the device, plug into the eth0 port and obtain an IP address. As you will see, the interface is very intuitive. Chapter 6 contains a more detailed, step-by-step set of installation instructions for both Pebble and m0n0wall.

Once into the configuration, set up the IP addresses for the local area network (LAN) and wide area network (WAN) ports. With these ports specified, you can change any of the firewall related features to match your environment. Once your environment is fully configured and tested, we are ready to move on to the captive portal.

NOTE...DIVERSE ENVIRONMENTS

While the example given in this chapter uses a separate AP, many users choose to embed a wireless card directly into m0n0wall. There are many ways to configure your network. The important thing is to have everything running and tested prior to moving to the next section. In this chapter, we describe a m0n0wall configuration using an external AP (a Cisco 1120). In Chapter 6, we describe a m0n0wall configuration using a wireless card in the Soekris.

Performing the Hack: Enabling Our Captive Portal

With our m0n0wall installed and ready to go, we can start captivating our users with the portal. The m0n0wall graphical interface makes this process quick and easy. Simply click the **Captive Portal** option under **Services | Captive Portal**, then perform the following:

1. Check the **Enable captive portal** option.
2. Click the **Browse...** button under **Portal page contents** and load the HTML for your captive portal. We will discuss formulating the language for your Terms of Service in the next section.

NOTE...RADIUS SERVERS

m0n0wall also supports Radius authentication. Simply add your server under the Radius Server section.

Figure 3.5 shows the Captive Portal tab. We will discuss the Pass-through MAC and Allowed IP Addresses tabs in a later section.

Figure 3.5 The Captive Portal Tab

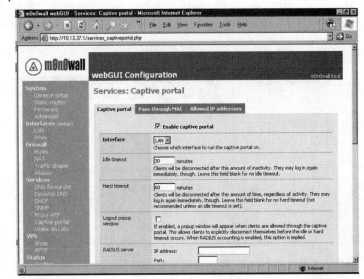

Writing Our Terms of Service

With our Captive portal up and running, the next step involves writing the legal ToS. This agreement educates users about their responsibilities when using the network, and protects the community network operators from legal liability. In this litigious world, even giving away wireless involves some risk. The ToS on our captive portal helps keep users informed and allows the network operators to focus more on building great wireless networks.

Here at SoCalFreeNet we use an agreement that basically says:

"Here is some free wireless. Don't abuse this, or we will take this away from you. Abide by the law and be nice to your fellow community members."

The ToS goes on to make some strong statements waiving liability for SoCalFreeNet and the hardware owner if a community member does something foolish like licking their access point while it's plugged in. While we realize there might be some odd rules in the legal system, no one wants to go to court over a foolish community user's mistakes. Some users simply make poor choices, and we don't want the network operators and community to suffer the consequences of their mistakes. Preventing these mistakes in the first place is a good approach, and SoCalFreeNet prides itself on making our wireless community as safe as possible. Building safe systems helps everyone enjoy their stroll through the community network park.

Our community agreement took a great deal of effort to formulate. Correctly building this document requires the help of legal experts. Unfortunately, legal expertise is foreign to the authors. We recommend you seek the assistance of a good technology lawyer when writing your own ToS. If hiring lawyers goes beyond your budget, we recommend contacting a local wireless users group, or ourselves (www.socalfreenet.org) and asking for help. By pooling our resources we can achieve critical mass and get things done.

Figure 3.6 shows the ToS screen users are presented with when they first attempt to browse to an Internet Web page. Notice the Web site in the *address bar* is www.wsj.com. Once the user agrees to the Terms of Service, they will automatically re-connect to the Web page they originally requested.

Figure 3.6 The ToS Screen

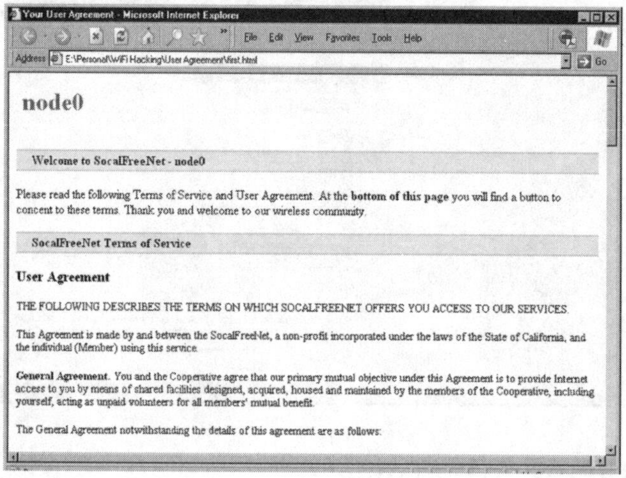

NEED TO KNOW…WRITING A PROPER ToS

The authors do not claim to be legal experts. Writing a proper User Agreement or Terms of Service requires expertise in technology law. The publisher and authors recommend seeking the appropriate legal help when constructing legal documents related to your own community network.

Captive Portal Graphics

With our captive portal fully operational, we can display a Web page to the user with our ToS. Generally, we develop the ToS page offline and then load the page on our m0n0wall device. However, loading a flat HTML page in m0n0wall limits our ability to include graphics. The graphics in the aforementioned example are flat and driven by simple color differences. To advance beyond this limitation requires a little tweak to our configuration and our portal page.

While the portal page provides a way to display text, m0n0wall does not allow us to upload local images (storage space is too limited on our CF card). Images make Web sites come to life and help bring dimension to our community. The SoCalFreeNet group logo makes a nice addition to our ToS page. Adding images requires a few simple steps:

1. Identify a Web server to host your graphics. Our graphics reside on the *Web/logs* server that sits outside (WAN) of the m0n0wall.

2. Once we have the images available on a Web site, we add an <IMG> tag to our portal page with a fully qualified reference to the image. This is a fancy way of telling us to use the whole Web address when referencing the image's location. Instead of using the normal <IMG SRC="/images/logo.gif"> we use <IMG SRC="http://172.16.1.186/images/logo.gif">. This tells the user's browser to go to a different server to get the image for our portal page. In our case, this is a Web server sitting inside our network.

NOTE...DNS ISSUES

We found that the captive portal does not allow DNS through before the user accepts the agreement. To address this, we use direct IP references instead of host names. If you don't know the IP address of the server you're trying to reference, use the ping utility to discover this information.

3. Configure your m0n0wall captive portal to allow users to the IP address of the server holding the images. To do this, click the **Services | Captive Portal** options on the right menu in the m0n0wall, then select the **Allowed IP addresses** tab just above the configuration items. Finally, click the **plus** icon on the far right in Figure 3.7; the server's IP address is 172.16.1.186. When referencing other servers, make sure you have the permission of the Web administrator to reference their site. This is considered good netizenship on the Web.

Figure 3.7 An Allowed IP Addresses Screen

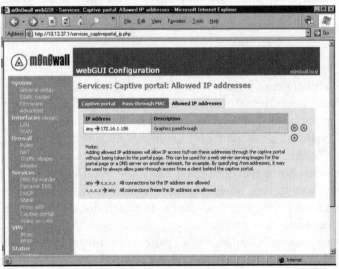

Figure 3.8 shows our enhanced ToS page with the SoCalFreeNet logo. This concept can be stretched even further with frames. Opening our mind to innovation often yields some interesting

results. Members of SoCalFreeNet continue to strive creatively to help make our community more enjoyable and safe.

Figure 3.8 Enhanced ToS Page with the SoCalFreeNet Logo

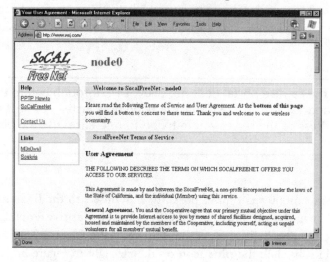

NEED TO KNOW...PERMISSION

Obtain the permission of Web administrators before linking directly to their site from within your captive portal.

Building a PPTP VPN

The realities and challenges of open wireless technologies lead to many interesting solutions. The emergence of Wired Equivalent Privacy (WEP) encryption offered an early form of protection. The integrity of this solution slowly dissipated as security analysts found vulnerabilities, and crackers began breaking the encryption. The weakness in this encryption mechanism gave rise to the Wi-Fi Protected Access (WPA) solution and later, 802.11. Now WPA faces some security scrutiny, as analysts start uncovering issues. Some speculate these issues will result in the fall of WPA. So with WPA on shaky ground, we chose to seek other solutions. While 802.1x technologies hold promise, this solution requires significant infrastructure to deploy. Furthermore, deploying this technology to old operating systems comes with some challenges.

Facing the encryption challenge requires patience and ingenuity. Many creative users turn to VPN technology to provide a higher level of protection. These encryption tunnels act as a cryptographic highway, protecting wireless traffic from outsiders trying to intercept the flow. The movement toward VPN technologies helps users enjoy the wonders of wireless. For our wireless VPN, we chose PPTP. While the PPTP protocol comes with issues, many of the security findings have been repaired.

For the issue surrounding password guessing attacks, we'll use long, complex passwords to mitigate much of this risk. While PPTP still presents some risk, we felt our solution offered a marked improvement over sending information across the community network in cleartext. Clearly, every wireless project must select the encryption tools that give them comfort. Many of our nodes do not use encryption at all and prefer an "open" approach. While this is considered less secure, it is easier to deploy and support. As with all security decisions, you must use judgment to balance between security needs and usability.

Note...Early Woes of PPTP

In 1998, the famous security analysts Schneier and Mudge released a pivotal paper (www.schneier.com/pptp.html) on weaknesses in Microsoft's PPTP implementation. Microsoft fixed many of these issues with subsequent releases (fixes require the DUN 1.3 upgrade); however, the password guessing threat still applies. Using complex passwords longer than 14 characters helps mitigate much of the risk from password guessing attacks.

Preparing for the Hack

Once we've gained some comfort from our VPN solution, we can jump into its implementation. Preparing for this hack follows the same requirements as discussed in the captive portal hack. Make sure to install your m0n0wall software, configure the ports, and test all connectivity. This helps ensure thatwe focus on the VPN configuration rather than addressing other configuration issues.

Warning: Hardware Harm

While our PPTP tunnel provides protection for users smart enough to use the tunnel, our network still allows users to connect using cleartext. Since many wireless users lack the security knowledge to understand the dangers of cleartext communications, we must actively educate users about good security habits and enforce their use.

Performing the Hack: Enabling the VPN

Configuration consists of two major phases. The first phase focuses on setting up the m0n0wall device, while the second focuses on configuring a client. Fortunately, the client comes embedded in many versions of Windows, including Windows 95, NT, 2000, XP, and higher versions.

On the server side, m0n0wall version 1.1 supports IPSec and PPTP tunnels. Setting up PPTP involves the following easy steps in m0n0wall's Web interface:

1. Select the **VPN | PPTP** option from the left menu bar. Notice both the Configuration tab and the Users tab near the top of the page. We will use both of these during this configuration.

2. In the **Configuration** tab, select the **Enable PPTP server** radio button.

3. In the **Server Address** box specify the IP address for the PPTP server. This is the IP address of the m0n0wall that clients will use once their VPN tunnels get connected. In our example, the server's address is set to 10.13.37.2. This IP address is simply one IP above our default gateway address on the LAN side. *Remember to choose an address that falls outside the DHCP range you specified on the* **Services | DHCP** *menu.*

4. Enter a network address in the **Remote address range** field. If you're unfamiliar with subnetting and how to choose an appropriate /28 network address, I recommend using a zero in the last octet. Our example uses the 192.168.13.0 range.

5. Optionally, you can select to enforce 128-bit encryption. We recommend using this unless you run into some conflict with a user.

6. Click the **Save** button at the bottom. We will get a message back at the top stating "The changes have been applied successfully." See Figure 3.9 for the final result. We now have a running PPTP server on the m0n0wall, and simply need to add some user accounts.

Figure 3.9 VPN:PPTP Configuration

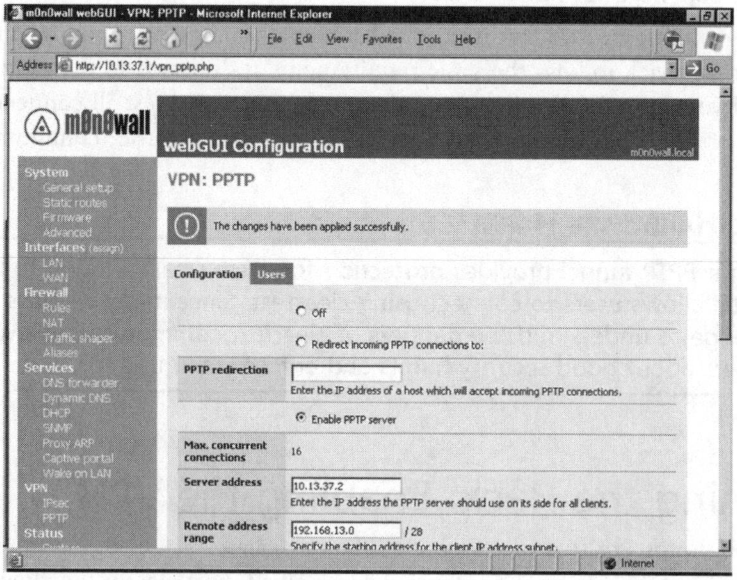

7. While still at the **VPN | PPTP** menu option, click the **Users** tab toward the top of the screen.

8. Select the **plus**-shaped icon on the right side of the screen and a new user screen will appear. Figure 3.10 shows the new user screen.

Figure 3.10 The New User Screen

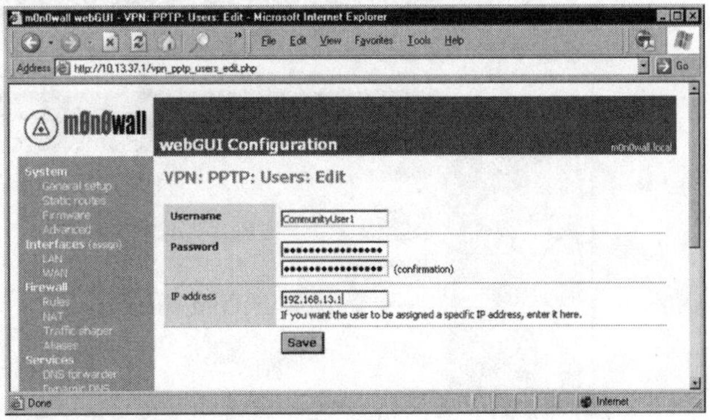

9. In the **Username** field enter the desired logon ID. In our example, we will use CommunityUser1.

10. For the **Password** field, enter a long and complex password twice. By long and complex, we mean use numbers, upper- and lowercase alpha characters, and special characters (like &). One handy memorization technique uses the first letter from each word of a song. For example (please don't use this example in your environment) "*Mary Had A Little Lamb, A Little Lamb, And Its Fleece Was White As Snow…*" would equate to a password of *2MhallAllAifwwas2*. We added the number 2 onto the front and back to increase the complexity.

NEED TO KNOW…M0N0WALL PASSWORDS

At publication of this book, m0n0wall did not support special characters (~!@$%^&*…). If a later version of m0n0wall supports this feature, we highly recommend using special characters. This increases the difficulty of password attacks exponentially.

11. In the **IP address** field, enter the IP this user will always get. We highly recommend using this option, since this will give us direct traceability from our logs to this PPTP user. This IP address must fall within the range we specified in the Remote Address Range on the first tab (192.168.13.0/28). In our example, we chose the first IP in this range (192.168.13.1).

12. For these changes to take effect we must click the **Apply changes** button at the top. Notice this will break all users currently connected through the PPTP VPN. Since we are setting our VPN up for the first time, this shouldn't be an issue. Once again, a message will appear at the top of the screen informing us that our changes have been applied. Figure 3.11 shows the final results.

Figure 3.11 The Final Results

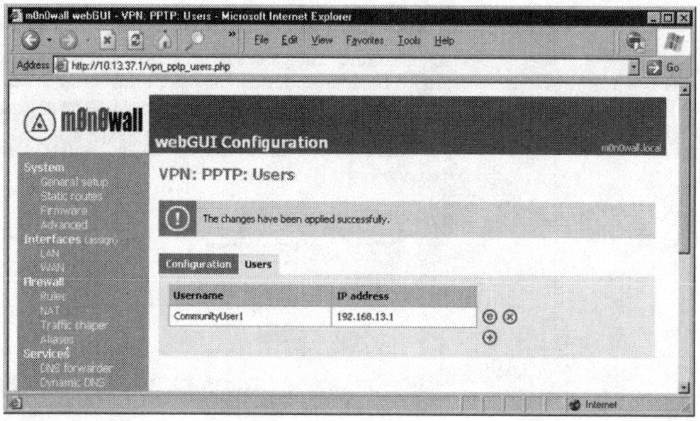

13. The final step on the m0n0wall configuration requires us to make a firewall rule to allow PPTP traffic to flow. Click the **Firewall | Rules** menu on the left. A configuration menu will appear with one rule already established on the *LAN interface*.

14. Once again, select the **plus** icon on the far right section of the configuration screen. There are a number of other icons here, including an *e* for edit and an *x* for delete, as well as arrows for moving rules up and down. m0n0wall processes firewall rules from top to bottom. Figure 3.12 shows this new rule screen.

Figure 3.12 The New Rule Screen

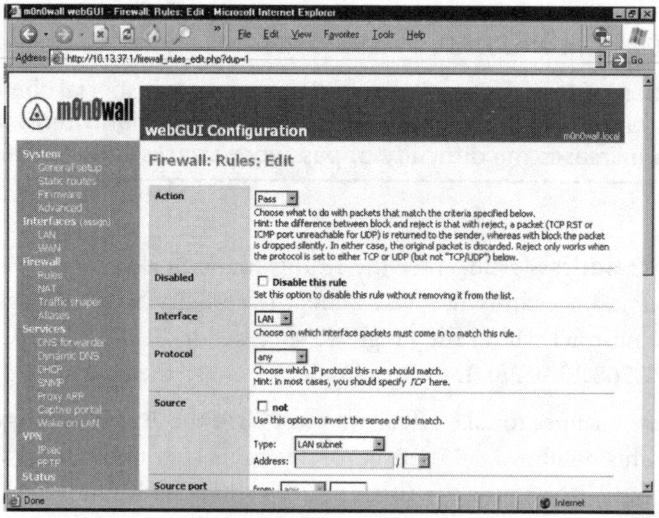

15. A new firewall rule screen will appear. In this screen, leave the **Action** field populated with **Pass**.

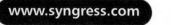

16.   Select the **PPTP** option from the **Interface** drop-down box.

17.   Under the **Protocol** drop-down box, select the **TCP/UDP** option.

18.   You can leave the rest of the fields to their defaults or modify them to fit your environment. In the **Description** field, enter a description of this rule. We use "PPTP clients -> internet" to remind ourselves of the role this rule fulfills.

19.   Click the **Save** button at the bottom of the screen. This will bring you back to the **Firewall | Rules** menu, which will display your new rule under a section called *PPTP clients*.

20.   Once again, we must click the **Apply Changes** button to put the new rule into effect. A message will appear confirming the changes have taken effect. Figure 3.13 shows the final results.

Figure 3.13 Firewall Rules, Final Results

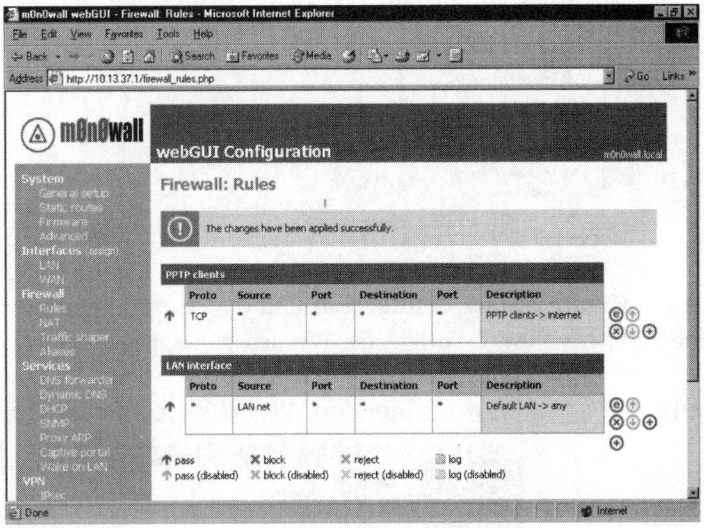

With our server fully configured, we are ready to move onto setting up the PPTP client on our community user's system.

NOTE...PASSWORD SECURITY

Choosing poor passwords can lead to the compromised security of your VPN. End users often dislike complex passwords. Try to be inventive in choosing passwords that the user can remember and still make good security sense.

Configuring Our Community Users

On the user's side, we will leverage the PPTP client already built into Windows. Our example will use Windows XP to demonstrate the setup. The configuration is similar between all versions, from Windows 95 up to the most recent version.

1. Click the **Start | Network Connections.** As seen in screen Figure 3.14, the Network Connections window will appear with your current network adapters already visible.

Figure 3.14 Clicking the Start | Network Connections Screen

2. Click the **Create a new connection** link on the upper-left side of the window. As seen in Figure 3.15, the **New Connection Wizard** will appear.

Figure 3.15 The New Connection Wizard Welcome Screen

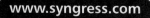

3. Click the **Next >** button, and the **Network Connection Type** dialog will appear. As seen in Figure 3.16, select the **Connect to the network at my workplace** radio button.

Figure 3.16 Selecting the Connect To The Network At My Workplace radio button

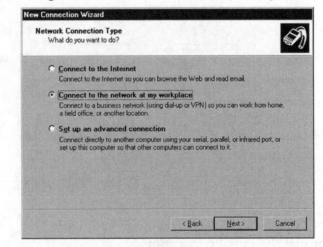

4. Clicking the **Next >** button brings up the **Network Connection** screen. As seen in Figure 3.17, select the **Virtual Private Network connection** option.

Figure 3.17 Selecting the Virtual Private Network Connection option

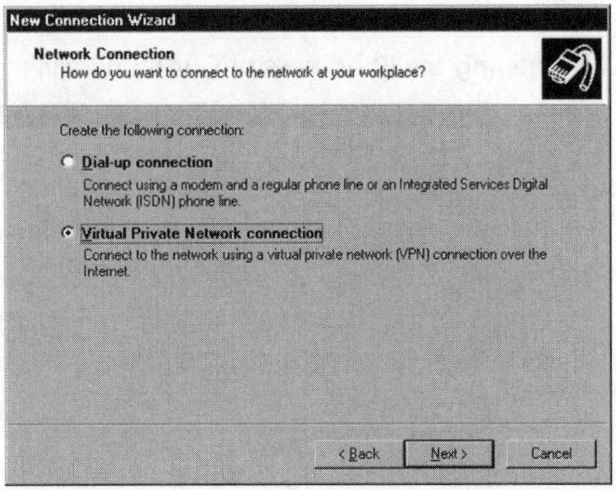

5. Clicking the **Next >** button brings up the **Company Name** prompt. Enter a description for your PPTP connection here. As seen in Figure 3.18, we entered PPTP to Community Wireless.

Figure 3.18 Description Entered for PPTP Connection

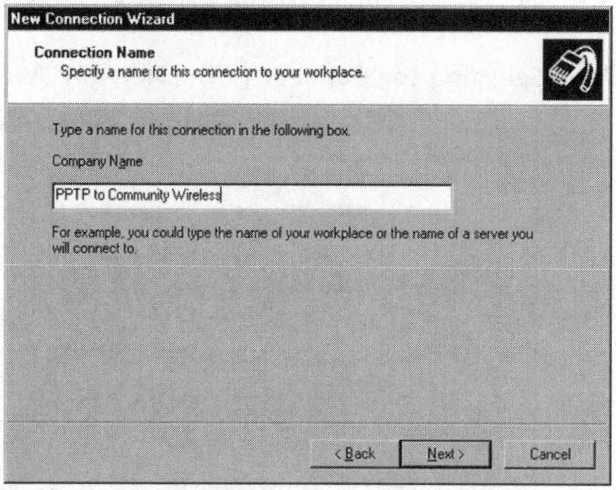

6. Clicking the **Next >** button *may* bring up an **Automatic Dial** dialog. If this occurs, select the **Do not dial the initial connection** option.

7. Clicking the **Next >** button brings up the **Server Name or Address** dialog. As seen in Figure 3.19, we entered the IP address of our m0n0wall. This is the address of the LAN interface on the m0n0wall. In our example, we used 10.13.37.1. (This address can be found in our m0n0wall by clicking the **Interfaces | LAN** menu item from the m0n0wall Web configuration.)

Figure 3.19 Entering the IP Address of Our m0n0wall

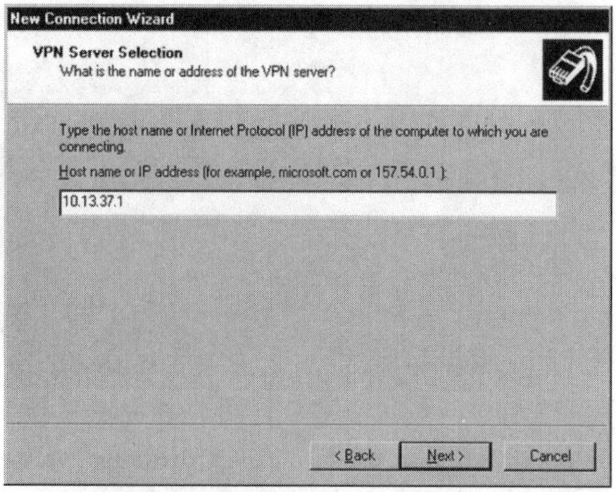

8. Clicking the **Next >** button brings up the **Create this connection for:** dialog. If we want all users to use this connection, select **Anyone's use**; otherwise, select the **My use only** option. This is largely left to the discretion of the community member using the PPTP connection.

9. Clicking the **Next >** button brings up the **Completing the New Connection Wizard**. Optionally, we can add a shortcut to the desktop.

10. Clicking the **Finish** button brings up a **PPTP authentication** box, as seen in Figure 3.20. To test our settings, enter the username we specified in the previous m0n0wall configuration.

Figure 3.20 The PPTP Authentication Box

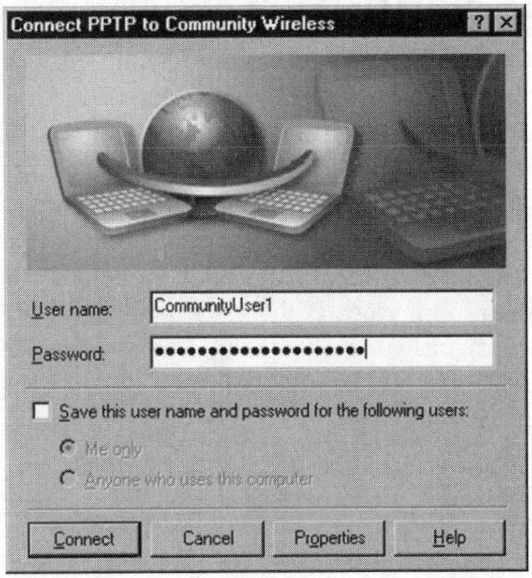

11. Clicking the **Connect** button brings up the **Connecting** dialog, as seen in Figure 3.21.

Figure 3.21 The Connecting Dialog Box

12. If we configured everything correctly, we will get a dialog box telling us we are registering on the network. Figure 3.22 shows this dialog box. Congratulations, our PPTP VPN is working!

Figure 3.22 Registering on the Network

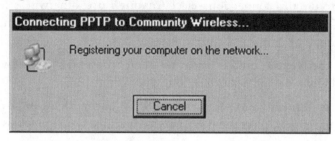

Hacking the Mind of a Wireless User

Good security starts with users. The community's users must take extra steps, like using good passwords, to make any of our optional security mechanisms work. All the mechanisms discussed up to this point focus on technology. In the world of computers, another skill known as social engineering comes into play. We define *social engineering* as the art of influencing people's actions through unconventional means. While social engineering is often associated with *black hats* or bad hackers, not all social engineering causes damage. Many of the same skills employed by black hats can be used to achieve positive results. The very concept of this book embodies this principle.

Hacking has taken on a bad persona as the news media hypes up cases of bad acts performed by hackers. The term hacker originally meant someone doing unconventional things to innovate and create new solutions. (Hence, the title of our book, *Wireless* Hacking). We too seek to push the technological "edge of the envelope" and help find new ways of creating a secure environment for our community network.

The word community implies social contact. This contact forms one of the fundamental ways for us to enjoy the community and share ideas. This channel offers us another pathway to promote security and educate our users on how to stay safe while enjoying the wireless park we create.

Preparing for the Hack

Building good supporting documentation helps users quickly learn to configure and manage their devices. With a limited volunteer force, the community network relies on friendly members taking the time to help others. Much of this help comes in the written form. Much like this book, our support documentation can help promote a strong community and good security.

Performing the Hack: The Beginning and the End

Strong security grows from a smart user base. Building this knowledge requires patience and a friendly demeanor. When a user approaches us with a question, we choose to think about the problem from their perspective and try to integrate their feedback into our network design. For example, we chose PPTP for the ease of implementation on the user's part.

As users start to understand wireless technology, and hear news about various wireless security components, they will grow curious. With well-developed help content, the user will have a place to research and learn. SoCalFreeNet uses the captive portal pages as a jumping off point for users to learn more about security and their role.

The user should take a few simple precautions when joining the community network:

- Always use a personal firewall. These firewalls often sit on the user's laptop or desktop. Windows XP comes with a newly enhanced firewall built-in.

- Use strong passwords to make password attacks more difficult.

- Even good passwords fall short sometimes. Many Web-based e-mail programs send the password through the network in cleartext. In our wireless network, this means other users might see a user's e-mail password. Users should make sure the little lock is sitting in the bottom-right corner of their browser when going to sensitive sites. This lock indicates the site uses SSL.

- Believe it or not, even the little lock isn't a full proof way to protect us. Some attacks use a man-in-the-middle device and can still see our encrypted traffic. For this reason, we encourage users to authenticate to the PPTP tunnel and make sure the lock appears as well.

- For highly sensitive browsing, consider doing this through more conventional means.

- Patch your systems on a regular basis. Users may want help understanding how to evaluate and implement patches.

- Teach your kids about the Internet and how to stay safe in the cyber world.

- If you get an uneasy feeling when browsing a Web site, stop and think about the security. Follow your instincts. The Internet mirrors life in many ways, and the cyber world has its own ghettos and undesirable areas. Avoid online merchants with poor reputations or poor-quality sites. They often treat security as non-essential.

While the SoCalFreeNet architects continue to seek and offer secure alternatives, the real security lies in the hands of the users. If users choose to ignore the security options we offer, our effort has gone to waste.

Our socialization of security into the community serves as the most fundamental element of good security. Making security important and easy for users yields the best results. If we use hacking to help our users learn security, we stand a better chance of securing our community network.

NEED TO KNOW...SECURITY AWARENESS

The list provided only covers the highlights of SoCalFreeNet's security awareness communication. We realize a great deal of material exists for helping users, and this list could grow many fold. The list provided serves as a sample.

Other Hacks

Your community network may come with other interesting challenges. Here are a few other ideas to consider when building a safe environment for your network:

- **Squid Proxy** An opensource tool called Squid Proxy can be employed to prevent users from getting to inappropriate sites. These sites might include pornography or hate sites. This tool is highly rated, and in addition it can help reduce the amount of traffic consumed on our broadband link through Website caching. www.squid-cache.org

- **Snort** Another opensource tool called Snort conducts intrusion detection. By scanning the traffic passing over the network, Snort can alert us to attacks coming from the wireless network. Recently, the Snort team added specialized functionality to help detect wireless attacks. www.snort.org

- **OpenSSH** Setting up an OpenSSH VPN. OpenSSH offers a feature called port forwarding. By using a non-interactive login with port forwarding, we can create a very nice VPN with security beyond our PPTP solution. If we have the infrastructure, a hierarchical mutually authenticated solution like EAP-PEAP offers maximum protection.

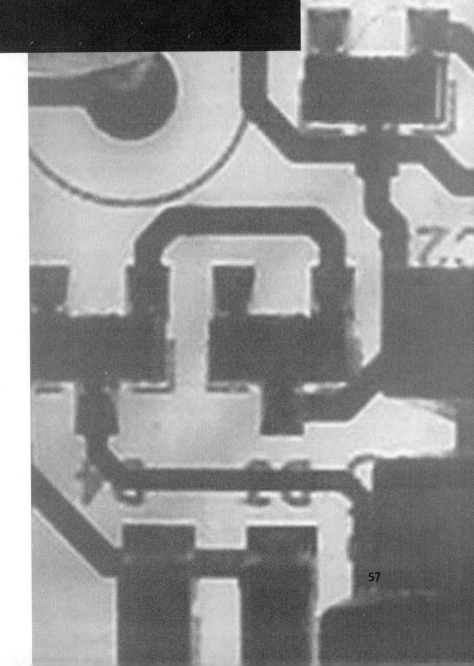

Part II

Hacking Projects

Chapter 4

Wireless Access Points

Topics in this Chapter:

- Wi-Fi Meets Linux: Linksys WRT54g
- Soekris Single-Board Computers
- Hacking a Proxim 8571

Introduction

In this chapter, we review wireless access point (AP) hardware options. Some APs are reflashed off the shelf; others are built using single-board computers and Linux. This chapter will serve as an introduction to all your hardware options.

Wi-Fi Meets Linux

Setting up Linux machines to act like wireless APs is certainly nothing new. Using HostAP drivers, Linux boxes can emulate the functionality of an AP in infrastructure mode and service wireless stations using off-the-shelf 802.11 gear.

Although it's possible to set up a large tower case running Linux as your AP, this method certainly has some disadvantages. First, the pure size and weight of the tower PC makes it somewhat difficult to mount in tight quarters where ample free space is lacking (such as ceiling crawl spaces, rooftop antenna masts, or the like). Second, a tower PC uses lots of power. Third, tower devices have a number of moving parts, such as power supply fans and hard drives. The more moving parts a device has, the higher the risk of a hardware failure. Fourth, tower PCs tend to be fairly expensive. Finally, tower PCs can be just plain ugly!

Of course, the advantages to running Linux on your AP are significant. Having a shell gives you an enormous amount of flexibility, compared to the restrictive Web-based management interface of your typical off-the-shelf consumer grade AP. With Linux, you have control over every aspect of the device's configuration and operation.

To take advantage of the benefits provided by Linux without the hassles of running desktop PCs, you can either reflash off-the-shelf APs or utilize a single-board computer (SBC) such as Soekris. With these devices, you have a small form factor, no moving parts, and low power consumption. The net result is an ultra-portable, ultra-reliable hardware device!

Reflashing

One of the earliest attempts to reflash a consumer-grade AP was the OpenAP project by Instant802 Networks. This project created a method whereby users could reflash a Eumitcom WL11000SA-N board (such as a US Robotics USR 2450, SMC 2652W or Addtron AWS-100) with a fully functioning Linux operating system. The drawback to the OpenAP method was that it required the AP case to be cracked open and a Static RAM card to be inserted for the reflashing. Static RAM cards were often expensive and difficult to obtain (this project was covered in detail in Chapter 14 of *Hardware Hacking: Have Fun While Voiding Your Warranty,* ISBN 1932266836, published by Syngress). For more information about the OpenAP project, please visit http://opensource.instant802.com/.

Linksys WRT54g

One of the most popular modern APs for Linux reflashing is the Linksys WRT54g. This device supports 802.11 b/g and has a built-in four-port switch. The native firmware supports WPA, NAT, DHCP firewall, and other functionality found in a standard AP. By default, the RF power output is rated at 18 dBM (63 mW). One of the nice features of the WRT54g is its size; it measures just 7.32" wide x 1.89" high x 6.89" deep and weighs just 17 ounces. It operates on 12V DC power (1A). Another major advantage of the WRT54g is the external RP-TNC antenna connectors.

Perhaps the only downside is the environment temperature rating, which is listed as 32 degrees, or 104 degrees F (0–40 C). This relatively limited range makes this device more suited for indoor deployments, except in areas with the mildest of weather. When operating in extreme temperatures, the device can become unreliable and "lock up," requiring frequent rebooting.

NEED TO KNOW... LINKSYS WRT54G HARDWARE SPECIFICATIONS

WAN port: One 10/100 RJ-45 port
LAN port: Four 10/100 RJ-45 ports
Channels: Eleven (USA), 13 (Europe), 14 (Japan)
LED Indicators (2.0): power, DMZ, WLAN, port 1/2/3/4, Internet
CPU: Broadcom BCM4702KPB 125 MHz (1.x), Broadcom BCM4712KPB 200 MHz (2.0)
RAM: Sixteen megabytes—IS42S16400 RAM Chips (Qty. 2)
Flash: Four megabytes—AMD AM29LV320DB-90EI (1.X), Intel TE28F320 C3 (2.0)
RF: Mini-PCI slot (1.0), integrated (1.1, 2.0),

Many different firmware distributions are available for the WRT54g. Keep in mind that different hardware versions of the WRT54g are in circulation, including 1.0, 1.1, and 2.0. Each version has slightly different hardware. For example, version 2.0 has an 18 dBm radio, whereas earlier versions had a 15 dBm radio. In this book, we will experiment with the hardware 2.0 version and will review products from:

- **Sveasoft** www.sveasoft.com

- **Newbroadcom** http://sourceforge.net/projects/newbroadcom

- **OpenWRT** http://openwrt.org

- **eWRT** www.portless.net/menu/ewrt

- **Wifi-box** http://sourceforge.net/projects/wifi-box

- **Batbox** www.batbox.org/wrt54g-linux.html

- **HyperWRT** www.hyperdrive.be/hyperwrt

Sveasoft

Unlike any of the other distributions discussed in this chapter, Sveasoft offers two versions of its firmware. The most current version (called a *pre-release*) is only available to subscribers who pay a $20 annual subscription fee. The company's "Current Stable/Public Release" is available free of charge. As of this writing, the pre-release is referred to as Alchemy, and the free version is referred to as Satori.

The first step to installing Satori is to download the firmware binaries. These are available from www.linksysinfo.org/modules.php?name=Downloads&d_op=viewdownload&cid=8. Upgrading your firmware is extremely simple and can be done via the native browser-based management interface.

Warning: Hardware Harm

 Anytime you are upgrading firmware, be sure to use a wired (not wireless) connection. Simply plug in a Cat5 Ethernet cable between your computer and one of the four switched ports of the WRT54g. Failure to use a wired connection increases the risk of a failed firmware update. If a firmware update fails in the middle of an upgrade procedure, you could damage your WRT54g router. A good Web site to learn about WRT54g recovery is http://voidmain.is-a-geek.net/redhat/wrt54g_revival.html. Also note that reflashing with unofficial firmware will void your warranty.

To upgrade your firmware on a stock WRT54g, perform the following:

1. Connect a Cat5 cable from your PC to the Linksys (on port 1-4, not the Internet port).

2. Open a browser and point it to 192.168.1.1. Figure 4.1 shows the popup window that you will see.

Figure 4.1 WRT54G Login Prompt

Connect to 192.168.1.1
WRT54G
User name:
Password:
☐ Remember my password
OK   Cancel

3. Leave the username blank and use the password **admin**.

4. Click the **Administration** tab, then the **Firmware Upgrade** tab, as shown in Figure 4.2.

Figure 4.2 Firmware Upgrade Tab

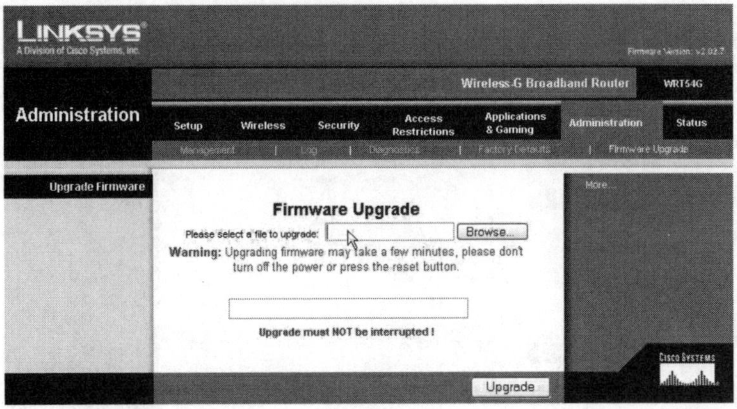

5. Click **Browse** and navigate to the Sveasoft file.

6. Next, click **Open** and then click **Upgrade**. Be sure not to interrupt power during the upgrade process. This upgrade could take several minutes and will result in a screen that should say "Upgrade is successful."

7. Click **Continue**.

You will notice that the look and feel of the Sveasoft management interface is identical to the stock Linksys interface. However, if you look in the upper-right corner, you will also notice that the firmware version is now being reported as Satori-4.0 v2.07.1.7sv. But that's just the beginning. The Sveasoft distribution includes dozens of new features not available in the stock Linksys firmware. For example, check out the **Administration | Management** tab. Figure 4.3 shows the stock Linksys firmware; Figure 4.4 shows part of the Sveasoft interface. Notice anything different? Clicking the **More** link on the right side of the screen (inside the blue area) reveals an upgraded help file that describes all the new functionality.

Figure 4.3 Stock Linksys Firmware

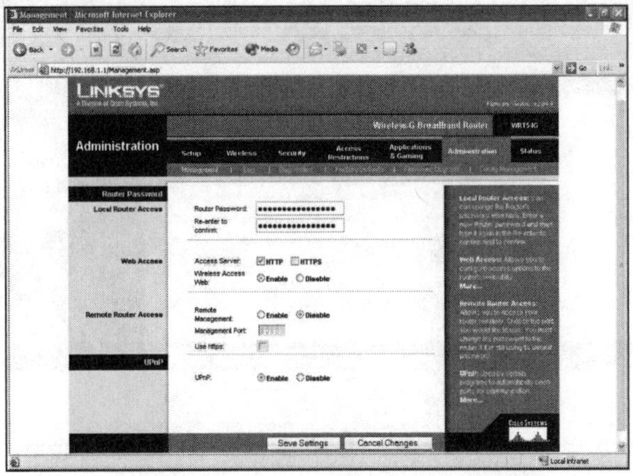

Figure 4.4 Sveasoft Interface

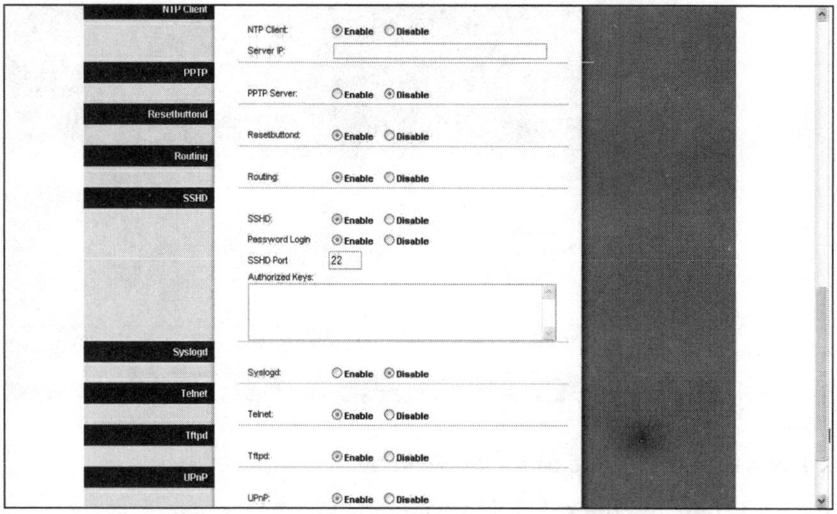

The Sveasoft firmware offers some impressive features. For example, you can now do things like modify the transmit power or change the antenna the device is using. Being able to select your antenna is an incredibly useful feature if you want to connect a high-gain omni to your Linksys instead of using the built-in diversity antennas. Sveasoft also introduces WDS to the Linksys device. Figures 4.5 and 4.6 show the standard Linksys options and the Sveasoft options, respectively for the Wireless | Advanced Wireless Settings tab.

Figure 4.5 Standard Linksys Options

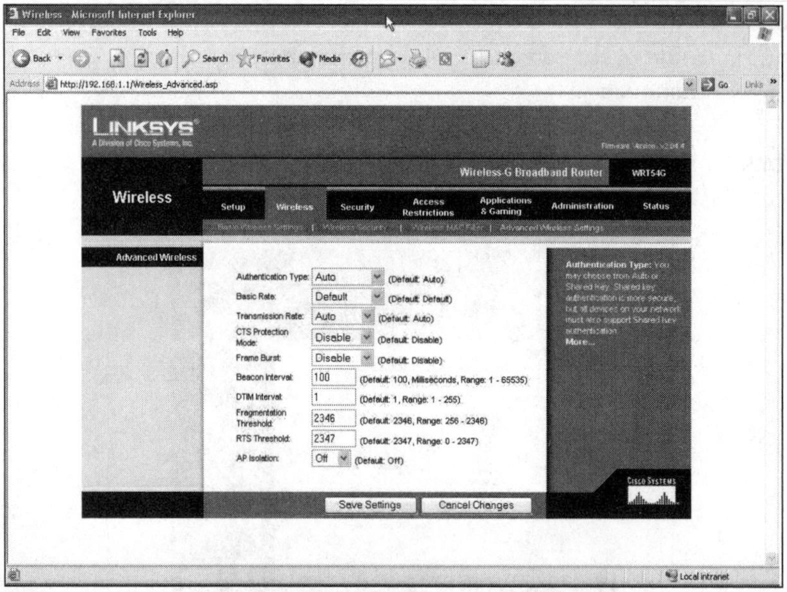

Figure 4.6 Sveasoft Options

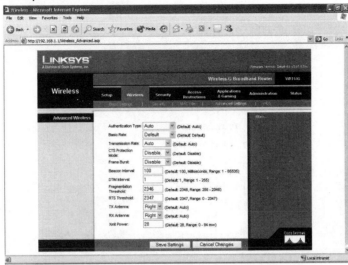

One of the fun features you can now enable is SSH. Here's how to set it up:

1. Navigate to the **Administration | Management** tab.

2. Scroll down to **SSHD** and select **Enable**.

3. Click **Save Settings** and **Continue**.

4. Now when you scroll back down to **SSHD**, you should see some new options.

5. Set Password Login to **Enable** and SSHD Port to **22**. Leave Authorized Keys blank.

6. Click **Save Settings** and **Continue**.

7. Now fire up your SSH client (such as putty.exe) and you can SSH to 192.168.1.1.

8. Log in with a username of **root**. Your password will be the router password (which, by default, is set to **admin**).

At this point, you are now SSH'd into your Linksys device! Just for fun, try typing **cat /proc/cpuinfo** and you should see the following output:

```
BusyBox v1.00-pre9 (1975.08.30-23:34+0000) Built-in shell (ash)
Enter 'help' for a list of built-in commands.

(none):[~]# cat /proc/cpuinfo
system type             : Broadcom BCM947XX
processor               : 0
cpu model               : BCM3302 V0.7
```

```
BogoMIPS              : 199.47
wait instruction      : no
microsecond timers    : yes
tlb_entries           : 32
extra interrupt vector : no
hardware watchpoint   : no
VCED exceptions       : not available
VCEI exceptions       : not available
dcache hits           : 2147418012
dcache misses         : 2012741550
icache hits           : 4294180837
icache misses         : 4215242303
instructions          : 0
(none):[~]#
```

If you have any difficulty using the Password Login feature, you can also create an SSH session using a public/private key combination. (This is also a much more secure method than usernames and passwords.) You can download a free copy of puttykeygen.exe to generate the keys from www.chiark.greenend.org.uk/~sgtatham/putty/download.html. Then:

1. Click **Generate** and follow the onscreen instructions.

2. Copy the Public key for pasting into OpenSSH authorized_keys file and paste it into the Sveasoft Management interface under **Authorized Keys**. Click **Save Settings** to save the key. Also, be sure to click **Save private key** and place it on your local hard drive.

3. Next, when you open **putty.exe**, click **Connection | SSH | Auth** and click the **Browse** key next to **Private key file for authentication**.

4. Find your private key file and click **Open**. Figure 4.7 shows the Putty configuration screen.

Figure 4.7 Putty Configuration Screen

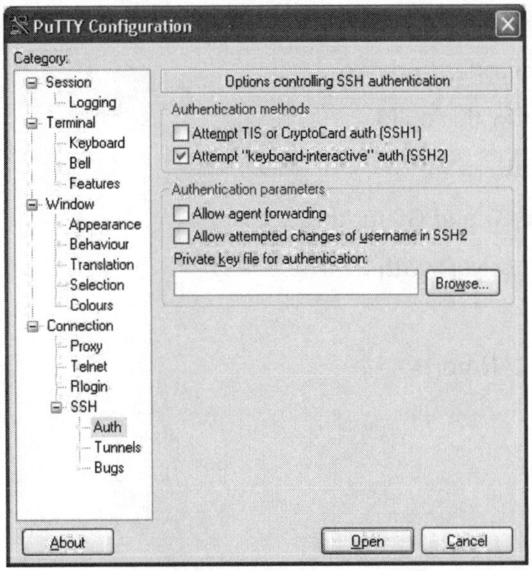

Have fun exploring the file system using SSH. You can **cat /etc/password** to look at the password file. If you want to see all the Web pages, you can type **cd /www** and then type **ls** to see all the files. To view the contents of any particular file, type **cat filename**. For larger files, type **cat filename | more**. For information about your Linux version, type **cat /proc/version**.

Here is a list of features for Satori 4.0, courtesy of linksysinfo.org:

The release adds the following features:

Auto channel select option

AP Watchdog timer option

New Management page help (thanks to Markus Baertschi)

SSH DSS keys now supported (thanks to Rod Whitby)

The following fixes were added:

NTP remote server field lengthened

Old port forwarding format supported

PPTP server fixed

webstr iptables filter fixed

adm6996 module moved

ifconfig broadcast addresses fixed

local dns fixed

Remote syslog fixed

rc_startup and rc_firewall fixes

Missing files and directories updated

Note: OpenSSL is not in this build

Standard Feature List

Compatible with both G and GS models

Linksys "AfterBurner" drivers with DMA

Setup

Default gateway for LAN ports

Advanced Routing

OSPF, BGP Routing

Wireless

Power mode selection

Antenna selection

Client mode (Ethernet bridging)

Adhoc mode

WDS peer-to-peer networking (10 links, multiple options)

Applications & Gaming

Modified to forward to any IP address

Administration

Bandwidth Management

Boot Wait

Cron

DHCP with static MAC->IP assignments

DNS Masq

Firewall control

Loopback option

NAS

NTP Client

PPP

PPTP VPN server

Resetbutton daemon

SSHD with public key or password login

Shorewall firewall

Syslog with remote logging

Telnet

Tftp

Diagnostics

Command Shell replaces ping and traceroute

Linux shell scripts rc_startup and rc_shutdown settable from the web

Status

Wireless signal strengths for clients, AP's, WDS links

Internal Modifications

BPAlogin fixes

Static DHCP leases

Added approximately 20 iptables filters

- include P2P, connection tracking

Added Quality of Service (for bandwidth mgmt)

Rewrote networking code for better stability

Added wireless connections daemon for client mode and WDS

Upgraded PPPD to 2.4.2

Added PPTP client and server

Various bug fixes to Linksys codebase

Latest Busybox

ADM6996 /proc interface

NewBroadcom

Because of the way that Sveasoft licenses its software, only subscribers who pay $20 per year can access the latest version of the firmware. However, due to subtle GPL licensing issues, a user may "fork" (branch off) the code and make modifications, then freely redistribute the "new" distribution. (However, they will no longer be eligible to receive updates via the subscription program.)

NewBroadcom 0.1 is a fork of the Sveasoft Alchemy-pre5.1 release. You can visit the NewBroadcom homepage at http://sourceforge.net/projects/newbroadcom. For more information about Sveasoft licensing issues, visit www.sveasoft.com/modules/phpBB2/viewtopic.php?t=3033.

Again, upgrading to NewBroadcom is done via the Web browser management interface. You can download the binary firmware file here: http://sourceforge.net/project/showfiles.php?group_id=115003, and follow the same instructions for upgrading firmware as found in the previous section (Administration tab | Firmware Upgrade tab). Alchemy-pre5.1 has the following added features compared to Satori:

- Multi-level QoS
- Ebtables 2.03
- Iptables 1.2.11
- Multiple client fix for client mode
- Optimization of kernel QoS code
- Busybox1.0–pre10
- Dropbear SSH 0.42
- 802.11x/WPA support on WDS interfaces
- Zebra OSPF with WDS interface support
- Squashfs V2.0 file system

HyperWRT

HyperWRT firmware works with WRT54g 1.0/1.1/2.0 and WRT54gs devices. You can download the binary firmware by visiting the HyperWRT homepage at www.hyperdrive.be/hyperwrt. At the time of this writing, the most current version was 1.3. Installing HyperWRT is as simple as using the management interface and upgrading the firmware, according to the previous instructions.

One of the nice features of HyperWRT is found in the Security | Firewall section of the management interface, where you can create filters and remove certain content from Web surfing activities. Supported filters include Java, ActiveX, proxies, and cookies. Another exciting feature, found under Status | Wireless, is the Site Survey option, which displays nearby APs. According to the HyperWRT site, these features already existed in the Linksys firmware but were intentionally disabled. HyperWRT simply enabled existing features that Linksys chose to remove. Figure 4.8 shows the filter options available.

Figure 4.8 Filter Options Available

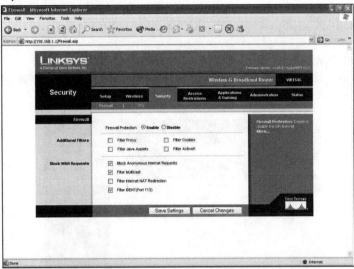

eWRT

eWRT is a firmware image that was forked from the Sveasoft Samadhi2 distribution. Samadhi2 is the version released prior to Satori (which is the version released prior to Alchemy). The eWRT homepage is located at www.portless.net/menu/ewrt. Like the other firmware options reviewed here, you can download the firmware binary and then easily upgrade your WRT54g using the browser-based management interface. Note that eWRT development is focused on the WRT54g v2 hardware platform. At this writing, the most current version of eWRT is 0.2 beta1.

One of the unique features of eWRT is the integration of NoCat, the captive portal system. A captive portal system works by redirecting the initial HTTP request to a specific Web site. Users must agree to terms of service before they are allowed to continue their browsing session. Figure 4.9 shows the portal screen users see when they first associate with the AP and try to navigate to a Web site.

Figure 4.9 Captive Portal Screen

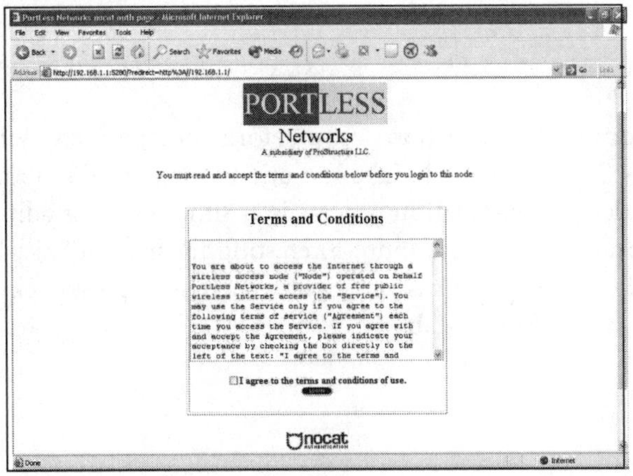

In the browser-based management interface, navigate to the **Administration** tab and then the **NoCat** tab, and you will find all the configuration options for the NoCat captive portal system. Figure 4.10 shows the NoCat options page in the management interface.

Figure 4.10 NoCat Options Page

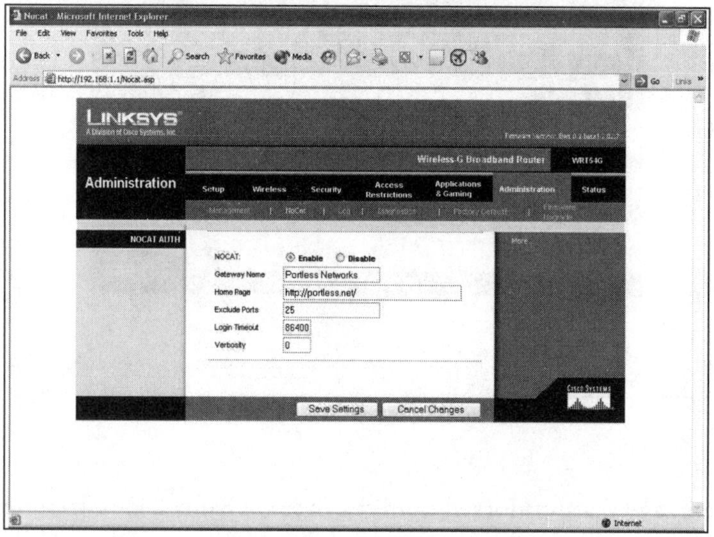

According to the eWRT homepage, the other key features include:

- NoCatSplash-based captive portal
- Traffic shaping with Wondershaper+iproute2
- SSH and Telnet management
- Wireless transmit power selection, client mode, Adhoc, WDS
- RSSI stats reporting for individual clients
- Remote syslogging

Wifi-box

The Wifi-box homepage can be found at http://sourceforge.net/projects/wifi-box. At this writing, the most current version is 2.02.2.1pre-wfb. Like the other firmware distributions mentioned so far, once Wifi-box is downloaded, the installation process is as simple as upgrading firmware via the browser management interface. One of the more interesting features of Wifi-box is a built-in PPTP VPN server. Figure 4.11 shows the VPN management screen, found by clicking the **Applications & Gaming** tab and then the **VPN Server** tab.

Figure 4.11 VPN Management Screen

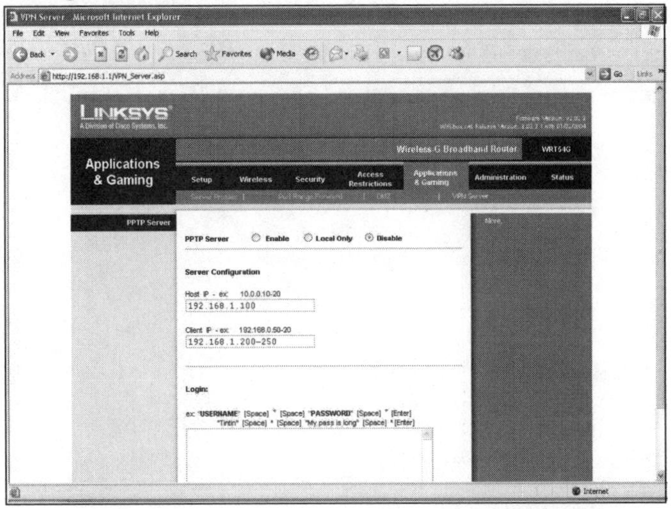

According to the release notes on www.linksysinfo.org, this version of Wifi-box includes:

First Release with 2.02.2 Source: (v2.02.2.1pre-wfb)

Supported: Wireless:

. Power Transmit Adjusting (12.75dBm ~19mW -> max 19.25dBm ~ 84mW)
. TX & RW Antenna Selection (Left Diversity Right)
. Support for 14 Channels (WorldWide)
+ Will support for Bridge and Repeater, WDS mode in final release

    System & Network:

. Support for subnet 255.255.0.0 & 255.0.0.0
. Static DHCP
. DNS Local
. SNMPD (Works right with mrtg)
. Support VPN Passthrough (IPSec - PPTP - L2TP)
. Add 'Server Profiles' for easy configure up to 14 Host Servers (
FTP,HTTP,HTTPS,DNS,SMTP,POP3,Telnet,IPSec,PPTP,Terminal,VNC,Emule,Ident,MSN)
. Up to 14 Port Range Forward settings
. VPN Server (PPTP) Built-in
. Support for Zone-Edit, Custom Dyndns DDNS
. Telnet Shell
. Remote Wake On Lan support

. Easy Reboot and Restart all service just a click

. Ping & Traceroute hacked for allow run shell command

. AutoRun Bash Script - Easy set an autorun script each time router reboot

. Status with more infos like Uptime & CPU Load, Wireless Client List

+ SSH Shell

+ Bandwidth Management

+ VPN Server IPSec

+ VPN Client (PPTP & IPSec)

** . = Current release | + = Will be add in next release **

All "hot & fresh" version:

iptables 1.2.9

PoPToP v1.1.3

pppd 2.4.2

busybox 1.0 pre7

pptp 1.4

net-snmp 5.1

Kernel 2.4.20 Tweaked

Batbox

Batbox is yet another mini-Linux distribution that operates on the WRT54g. It was written by Jim Buzbee. What's interesting about this distribution is that it resides only in RAM. If you power cycle your device, the changes are gone and you are left with your original firmware. The upside is that if you make any mistakes, you can turn off your device and when you power it back up, it's as good as new. The downside, of course, is that you lose all your changes whenever you power cycle the device.

Unlike the other distributions, the Batbox installation is not as simple as uploading new firmware. Rather, you must follow the lengthy instructions in the README file to install Batbox. To download the files, visit www.batbox.org/wrt54g-linux.html. At this writing, the most current version of Batbox is 0.51. Be sure that the Internet (WAN) port of the Linksys has been configured properly using the native firmware before you attempt Batbox installation.

If you're using Linux, you need to modify the wrt54g.sh script and include your linksys IP address and password. The IP address is located on the third line and by default is set to 192.168.1.1. The password is located on the 11th line. If you forget to modify wrt54g.sh with your password, the script is smart enough to prompt you for it before attempting Batbox installation.

If you're using Windows, the install process is a bit more involved. You will need to download and install CygWin from http://cygwin.com. Be sure to include wget from the Web package as well as ttcp from the net package. Uncomment the wget lines in wrt54g.sh. The wget lines are located on lines 54 and 55. When you open a CygWin window (shell), go to the directory where Batbox was unpacked (**cd /directoryname**). Next, make a copy of ttcp.exe and put it in the current directory (**cp /usr/bin/ttcp.exe** .). Finally, execute the script by typing **bash wrt54g.sh**.

If you receive an error message that looks like the following:

```
ttcp-t: socket
ttcp-t: connect: Connection refused
erron-111
ttcp error, status is 1
```

it is likely that your Internet (WAN) port is not properly configured.

This happens if, for example, you are configured for DHCP but do not have a Cat5 cable connected to the Internet port. If you are installing Batbox temporarily, you can simply set a Static IP address for the Internet Connection Type in the Setup | Basic Setup tab of the standard Linksys management interface.

Once Batbox is installed, you can simply Telnet to the box (no password is required) and you are automatically logged in as *root*. Obviously, this presents some security issues. In addition, Secure Shell (SSH) is not available. Alternatively, you can access a special Web server listening on port 8000 (http://192.168.1.1:8000). Note that the "standard" browser interface is still available on port 80 (http://192.168.1.1).

Keep in mind that the Batbox distribution resides strictly in RAM. Therefore, a quick reboot will restore your original Linksys firmware.

OpenWRT

OpenWRT is another Linux distribution custom-designed for the WRT54g. It was intended to be a very lightweight distribution and to offer very limited functionality but extensive support for adding features. This gives users the ability to custom-build their distribution and only add the features that they need, thereby saving precious system resources. Similar to Batbox, OpenWRT does not have a simple binary firmware file that you can simply upgrade via the browser management interface. Although the process for installation is not trivial, luckily you can find an excellent user guide and other documentation online. For more information on OpenWRT, visit http://openwrt.org.

WRT54G Shortcomings

Of course, one of the problems with reflashing a consumer-grade AP is that even with its newly enriched Linux firmware, the underlying hardware is still a consumer-grade AP! This means that the power output is typically low, and you might encounter a number of stability and reliability issues. Consumer-grade APs typically need to be reset (power cycled) more frequently then enterprise models. Some models have been known to require rebooting as often as once a week.

If you're looking for a more reliable hardware solution than a consumer-grade AP, your best bet is to look at an SBC. In the next section, we review SBC devices from Soekris Engineering.

Soekris Single-Board Computers

Soekris computers are extremely elegant SBC devices. The advantages of using a Soekris box include:

- Small form factor

- Lightweight

- Low power consumption

- No fans, drives, or other moving parts

- Ultra-quiet and reliable

In an abstract sense, a Soekris box functions just like a regular 486 PC. However, a regular PC would be difficult to mount in a weatherproof outdoor enclosure. A standard PC would also be more likely to require maintenance because of the power supply, hard drive, and other moving parts. Using a Soekris box is a great way to deploy an AP. In this chapter we'll tell you more about some of the hardware specifications for the Soekris line. In Chapter 6, "Wireless Operating Systems," you'll learn how to install m0n0wall and Pebble on these boxes.

NOTE

Various versions of Linux, OpenBSD, FreeBSD, NetBSD, and others can be installed on Soekris hardware, which is designed specifically for running open source software.

net4501

Originally, many Soekris devices were used as small, portable firewall systems. The net4501-30 is based on a 486-compatible CPU (an AMD ElanSC520) running at 133 MHz. It includes 64 MB RAM, three Ethernet ports, two serial (console) ports, a Compact Flash slot, one Mini-PCI slot, and a 3.3V

Figure 4.12 net4501 without a Case

PCI slot. The size of the board (without case) is a mere 4.85" x 5.7" and supports an operating temperature of 0–60°C. Figure 4.12 shows a photo of a net4501 without a case. Figure 4.13 shows a photo of a net4501 in a case, and Figure 4.14 shows a net4501 without a cover.

Figure 4.13 net4501 in a Case

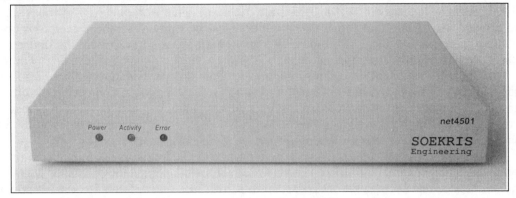

Figure 4.14 net4501 Without a Cover

net4511

The net4511-30 is also based on a 486-compatible CPU (an AMD ElanSC520), but it operates at 100 MHz. It includes 64 MB RAM, two Ethernet ports, one serial (console) port, a Compact Flash slot, one Mini-PCI slot, and one PCMCIA slot. The net4511 supports Power over Ethernet (PoE) using the IEEE 802.3af standard.

The size of the net4511 board (not including the case) is a mere 6.7" x 5.7" and supports an operating temperature range of 0–60°C.

NOTE: POWER OVER ETHERNET

When you set up an AP, there are generally two interfaces that need to be connected to the AP. First, you need to have an Ethernet cable to deliver the bandwidth and transfer the data. Second, you need to provide power to the device. The idea of Power over Ethernet (PoE) is simply to utilize the unused wires in a Cat5 cable to provide low-voltage power over the Ethernet cable. It turns out that a Cat5 cable has eight wires, and only four of them are used to provide data. (Note that Gigabit Ethernet does use all eight wires and therefore is not PoE compatible.) The IEEE 802.3af standard defines how to provide power over the unused wires. The net result: fewer cables!

In addition, using PoE, you can simplify your AP deployments and lower your implementation costs. By supplying power over the existing Cat5 cable, you also get more flexibility in terms of AP placement. Now you can put your AP anywhere you like, not just in places where a power outlet is available. This is particularly important when you are mounting APs on a rooftop, where power outlets might not be readily available!

Although it is possible to build your own homebrew PoE devices (www.nycwireless.net/poe), we highly recommend using a commercial version for safety and reliability. Many vendors offer PoE injectors (to introduce power into the Cat5) and splitters (to extract power from the Cat5). Note that sometimes splitters are also referred to as *taps* or *pickers*. If your AP supports PoE directly, you just need to purchase an injector. If your AP does not support PoE (and only accepts power via DC), you will need both an injector and a splitter.

net4521

The net4521-30 is very similar to the net4511 except that the net4521 has 2 PCMCIA slots instead of one. To accommodate the second PCMCIA slot, the board is slightly larger (but still very small), measuring 9.2" x 5.7". It also sports a slightly faster CPU, clocking in at 133 MHz. Figure 4.15 shows a net4521 board.

Figure 4.15 net4521 Board

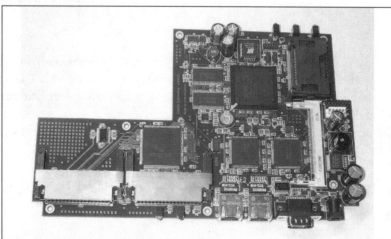

net4526

With the growing popularity of using Soekris devices as APs, Soren Soekris (the engineer who created these wonderful devices) developed a new product, the net4526, that hit a sweet spot for wireless applications. Two models are available: the net4526-20 and the net4526-30. The net4526-20 has a 100 MHz CPU (AMD ElanSC520), 32 MB of RAM, and 16 MB of on-board Compact Flash. The net4526-30 has a 133 MHz CPU (AMD ElanSC520), 64 MB of RAM, and 64 MB of on-board Compact Flash. Both models feature one Ethernet port, one serial (console) port, two Mini-PCI slots, and 802.3af PoE support. Amazingly, the net4526 board measures a tiny 4.0" x 5.2"! Like other models, the net4526 also has an operating temperature range of 0–60°C. As the Mini-PCI interface grows in popularity for 802.11 cards, this device is ideal for use as a wireless router. Keep in mind that the net4526 does not include a Compact Flash slot (the CF memory is built into the board). Therefore, be sure to select the model with enough CF memory for your application. The advantage of integrated CF memory is that you don't need to spend extra money for a CF card, and it allows the board size to be slightly smaller. The disadvantage of integrated CF memory is that it is difficult to increase available CF memory down the road. Figure 4.16 shows the net4526 board.

Figure 4.16 net4526 Board

net4801

Although it is not typically used as a wireless device, we include the net4801 in this chapter for the sake of completeness and because it is such a unique and interesting product. The net4801 is more of a midrange board, since it boasts a powerful CPU (a 266 MHz 586-class NSC SC1100) and is much more powerful than necessary for basic Wi-Fi applications. It also has 128 MB of RAM, three Ethernet ports, two serial (console) ports, a USB connector, a Compact Flash slot, a 44-pin IDE connector (for adding a hard drive), a Mini-PCI slot, and a 3.3V PCI slot. That is an amazing amount of power for a board measuring 5.2" x 5.7", with an operating temperature range of 0–60°C. Although it is clearly overkill for making a simple AP, this device is ideal as a communications appliance for other needs, such as a faster firewall, router, caching server, etc.

Soekris Accessories

Soekris conveniently offers packages for its SBC products. You can purchase them as standalone boards or as a package with the matching green case. Power supplies are also sold separately; you can choose between standard AC "wall wort" (adapter) options or 802.3af compliant PoE options. Soekris also sells PCI and Mini-PCI hardware accelerator encryption cards. The cards work perfectly as VPN accelerators or for scenarios where AES encryption is required. The vpn1401 (PCI) and vpn1411 (mini-PCI) support throughput of up to 250 Mbps. For more information, visit www.soekris.com. Figure 4.17 shows the vpn1401; Figure 4.18 shows the vpn1411.

Figure 4.17 vpn1401

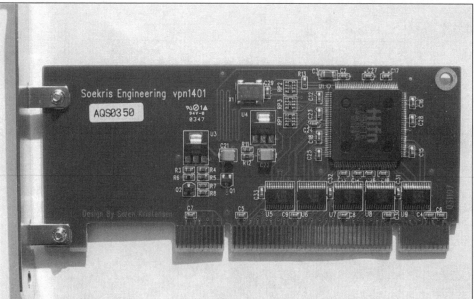

Figure 4.18 vpn1411

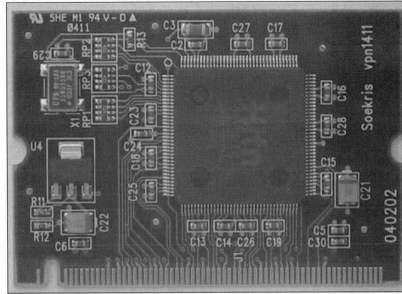

NEED TO KNOW... METRIX COMMUNICATIONS

If you're looking for a quick and easy way to deploy a Soekris device in an outdoor enclosure, a wonderful solution has been created by Metrix Communications. Its package integrates Soekris devices (a net4526) with a NEMA 4x weatherproof outdoor box. With Metrix, you now have a one-stop shop for all your outdoor AP kit needs! For more information, visit www.metrix.net.

Proxim 8571 802.11a Access Point

To create the 5 GHz backhaul links described in Chapter 2, one of the ways we were able to keep our costs low was to hack a Proxim 8571 802.11a AP. It turns out that cracking open the case of this device reveals a perfectly good PCMCIA 802.11a card, which you can install in a Soekris device and use as an 802.11 client with Pebble. (See Chapter 6 for more information about installing Pebble.) Figure 4.19 shows the back of an 8571 and its dual SMA female connectors.

Figure 4.19 Back of 8571

What makes this card special is that it has detachable antenna connectors. Being able to connect an external antenna is a feature not commonly found on 802.11a devices, but it's extremely useful for community wireless networks. By connecting an external antenna, this card provides a great deal of flexibility and wireless deployment options. Figure 4.20 reveals the main board with the lid removed. On the other side of this board is the PCMCIA card, shown in Figure 4.21. When connecting external antennas, use extreme care to ensure that your links are FCC compliant.

Figure 4.20 Main Board with Lid Removed

Figure 4.21 Main Board with PCMCIA Card

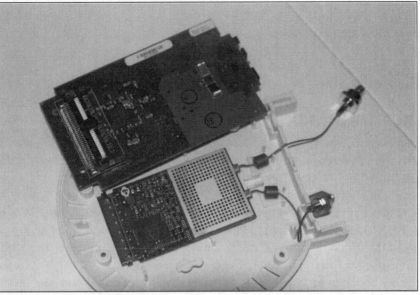

WARNING: HARDWARE HARM

This hack will void your warranty. Furthermore, whenever you open your case, you run the risk of accidental damage to your hardware. Use caution when performing this hack. Be especially careful when inserting a Proxim PCMCIA card into a Soekris or other PC device, because the casing of the PCMCIA card is not complete and can easily bend the pins of your PCMCIA socket if not aligned perfectly.

Preparing for the Hack

The primary advantage of the Proxim 8571 is low cost. In most cases, 802.11a gear is more expensive than 802.11g or 802.11b equipment because of critical mass and economies of scale. However, under some rare circumstances, it is possible to acquire 802.11a gear at extremely low costs because a model is discontinued or an aftermarket reseller is trying to clear out a large lot. You can find the 8571 as low as $20 or $30 by using price search engines such as www.pricegrabber.com or auction sites such as eBay.

The 8571's features include:

- Detachable antenna connectors (SMA-female)

- Support for three channels in U–NII 2: 56 (5.280 GHz), 60 (5.300 GHz) and 64 (5.320 GHz)

- SNMP support (disabled by default; see http://support.proxim.com/cgi-bin/proxim.cfg/php/enduser/std_adp.php?p_faqid=1147 for MIB information)

- A Telnet Interface (more about this in a moment)

The tools and materials required for this hack are as follows:

- Tamper-proof Torx screwdriver

Performing the Hack

The Proxim utilizes tamper-proof Torx fasteners on the 8571 case. The screw heads contain a six-sided star pattern with a center pin. To remove the center pin, you can drill it out, use a screwdriver to break the pin, or purchase a special tool.

The 33 piece security bit set (part number 2930) sold by Boston Industrial contains the bit you need to open the case without damaging the screw (www.bostonindustrial.com/bostonindustrial/2930.asp). For $10.95 plus $7 shipping and handling, you can get a complete kit with a variety of bits. It turns out that the only one you need to open the Proxim is the Star Drive: SD-8. (Note that the bit itself is labeled T-8H). However, if you are an avid wireless hacker, it is likely that the other bits will come in handy for other projects. Alternatively, you could do some searching online to try to find the individual bit required for this project. Remember that these are drill bits, not an actual screwdriver, so you will need a drill (or a screwdriver adapter that accepts drill bits).

To create a 5 GHz point-to-multipoint link, you will need a minimum of two Proxim 8571s. The first will be the AP. The second (and beyond) will be the client. For the AP, you can mount the 8571 in a Tupperware container (for short-term testing purposes), as shown with the lid removed in Figure 4.22 and in close-up in Figure 4.23.

Figure 4.22 8571 with the Lid Removed

Figure 4.23 8571 Close-up View

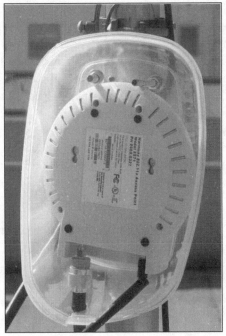

For longer-term deployments, the device should be mounted in a suitable outdoor enclosure. (More on outdoor enclosures can be found in Chapter 11.) Figure 4.24 shows an 8571 mounted in an electrical box purchased from Home Depot. Note that the sides of the case (which did not contain any electronics!) needed to be "shaved down" to accommodate the smaller box size.

Figure 4.24 8571 Mounted in an Electrical Box

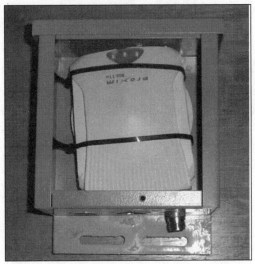

For configuration purposes, you can attach to the device using an 802.11a wireless client or by connecting an Ethernet cable directly to the 8571 device. One of the clever features of the 8571 is that it ships, by default, with an IP address of 169.254.x.x statically assigned to the wired Ethernet port. (The back of the device has a sticker with the actual IP address printed on it.) This way, when you connect a PC to the 8571 and it fails when attempting to lease a DHCP address, your PC will auto-assign itself an IP address in the 169.254.x.x range and therefore will automatically be able to "see" the Proxim. To view the management interface, simply open a browser and point it to the IP address of the 8571. When you're prompted for a username and password, the default is to leave the username blank and enter a password of **default**. By clicking the **Configure** and **Radio** tabs, you can set the channel, SSID, WEP, and other wireless parameters. Figure 4.25 shows a screen shot of the AP Radio Configuration options. Also note that in the Configure and Network tabs, you can enable or disable SNMP.

Figure 4.25 The AP Radio Configuration Options

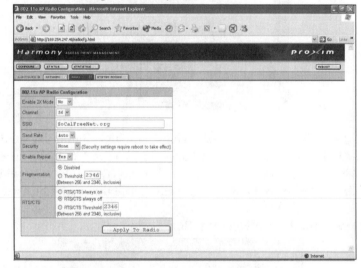

For the client side of the bridge, you should remove the 802.11a PCMCIA card and insert it into a Soekris or other suitable PC. The remainder of the 8571 device is no longer needed but should be saved as spare parts for the AP. In this manner, the 8571 is simply "harvested" for its client radio. Given the amazingly low cost of aftermarket Proxim 8571 devices, this method is the most cost-efficient way to build an 802.11a link. Figure 4.26 shows a close-up shot of a PCMCIA card after removal from an 8571.

Figure 4.26 Close-up Shot of a PCMCIA Card after Removal from an 8571

Note that the 8571 uses an Atheros-based radio. To configure your Soekris/Pebble device, per-form the following steps:

1. Enter the following:

   ```
   /usr/local/sbin/remountrw
   ```

2. Next, edit the /etc/network/interfaces file by typing:

   ```
   vi /etc/networks/interfaces
   ```

3. Comment out any lines in that file and replace it with the following:

   ```
   auto lo
   iface lo inet loopback

   auto ath0
   iface ath0 inet static
           address #insert IP address for your 802.11 card, i.e. 10.0.0.2
           netmask 255.255.255.0
           broadcast 10.0.0.255
           gateway 10.0.0.1
           up iwconfig ath0 ap #enter the MAC Address of the 802.11a AP on the other
   side of the link, i.e. 00:20:A6:47:f7:30
   ```

```
# alternatively use the following line (uncomment) if you want the client to look
# for a particular SSID instead of a specific AP MAC Address
#        up iwconfig ath0 mode managed essid socalfreenet.org

auto eth0
iface eth0 inet static
        address #insert IP address for your wired Ethernet port, i.e. 192.168.1.1
        netmask 255.255.255.0
        broadcast 192.168.1.255
```

4. To save your changes in the editor, press **Shift** and type **ZZ**.

5. Next, you will need to modify /etc/modules. (Again, type **vi /etc/modules**.) Add the line:

```
ath_pci
```

NOTE

If you have a Soekris device that supports a second Wi-Fi radio, you can use an 802.11b card and have one device operate as both an 802.11a backhaul and 802.11b client access radio. If you are using an 802.11b Mini-PCI card, you should add the line **hostap_pci** to the /etc/modules file. If you are using an 802.11b PCMCIA card, you can omit that step.

6. Next, don't forget to define the 802.11b radio (wlan0) in the /etc/network/interfaces file. For example:

```
auto wlan0
iface wlan0 inet static
        address 192.168.2.1
        netmask 255.255.255.0
        broadcast 192.168.2.255
up iwconfig wlan0 essid socalfreenet.org channel 1
```

7. Finally, to save your changes and reboot, enter the command:

```
/usr/local/sbin/fastreboot
```

Figure 4.27 shows an example of a Soekris box with a "harvested" 802.11a PCMCIA card, next to an 802.11b PCMCIA card. When selecting antennas, keep in mind that the 8571 AP operates in the U-NII 2 middle band (5.25-5.35 GHz). Again, always be sure to select antennas that are in compliance with FCC rules (or whichever rules apply in your country).

Figure 4.27 An Example of a Soekris Box with 802.11a and 802.11b Radios

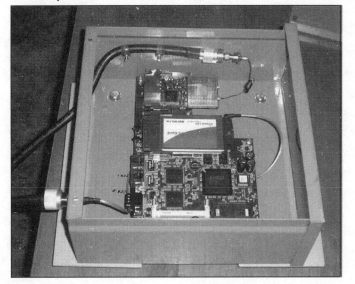

Under the Hood: How the Hack Works

You can learn more about the Proxim 8571 at the www.proxim.com Web site. Of particular interest is the April 2002 press release announcing the 8571 at www.proxim.com/about/pressroom/pressrelease/pr2002-04-01.html, which reads "The Harmony 802.11a Access Point—connectorized version (Model Number 8571) is available immediately for $695." You can also read the User Manual at www.proxim.com/support/all/harmony/manuals/pdf/857xman01.pdf. In addition, be sure to upgrade the firmware to the most recent version here: http://support.proxim.com/cgi-bin/proxim.cfg/php/enduser/std_adp.php?p_faqid=1227. Use the option **For stand-alone APs (no AP Controller)**.

If you are curious, the antenna connectors on the PCMCIA card are Radiall UMP series. You can find more information here: www.firstsourceinc.com/PDFs/ump.pdf. Furthermore, the Proxim 8571 does support PoE, but since it predates any IEEE PoE standards, the 8571 is not 802.3af compliant. For PoE operation, you should use a Proxim Harmony Power System, Model 7562. These can also be found at aftermarket resellers and auction sites. For more information, see the User's Guide at www.proxim.com/support/all/harmony/manuals/pdf/7562newmanb.pdf.

A quick port scan of the 8571 reveals two open TCP ports (80/HTTP and 23/Telnet) as well as one open UDP port (161/SNMP). Ahah! A Telnet port. Thanks to an anonymous poster on our Web site, you can now Telnet to the 8571 using the password *notbrando* and gain access to a special DebugTerm mode. Pressing the question mark (**?**) reveals the following list of commands:

```
Password->notbrando

DebugTerm->?
A = MAL registers
a = Atheros Radio Menu
b = netbuf debug
c = crash-o-matic
d = bridge tables
E = enet chip info
e = packet debug
f = radio tests
g = toggle watchdog
L = lock guided mode
l = enable debug log
M = mfg info
m = miniap info
n = net stats
o = reboot
p = print auth filtering stats
Q = quit
r = show radio settings
R = remote AP debug
s = show stacks
T = disable telnet
u = mem debug
v = version
V = display Config
w = write config
X = nuke config
Y = nuke image
z = write new bootrom
Z = write new image
0 = reset debug stats
1 = force deregister
8 = show 802.1x menu
Main->
```

Pressing the letter "r" (lower case) reveals interesting radio statistics.

```
Main->r


Radio State Down 100   resetOn = 0


Radio Misc Statistics
```

curTxQ =	0	maxTxQ =	1	curRxQ =	400	minRxQ =	0
txDescC=	0	TxPend =	0	rxDescC =	400	sibAge =	0
StaInPS=	0	StaDim =	0	psChange=	0	txUrn =	0
curtxPS=	0	maxtxPS=	0	PSQueue =	0	PSDeque=	0
curAltQ=	0	maxAltQ=	0	AltQueue=	0	AltDequ=	0
Rx =	0	Tx =	472	RxBad =	0	TxBad =	0
RxGood =	0	TxGood =	472	RxUni =	0	TxUni =	0
RxMulti=	0	TxMulti=	472	RxMgt =	0	TxMgt =	0
RxCtrl =	0	TxCtrl =	0	RxDscrd =	0	TxDscrd=	29
RuBrdg =	0	TuBrdg =	0	RmBrdg =	0	TmBrdg =	472
RepUnPk=	0	RepMuPk=	0	nullPtr =	0	hwReset=	0

```
802.11a settings
SSID- socalfreenet.org
Channel- 56
Main->
```

Pressing the letter "V" (upper case) displays some interesting Configuration data:

```
Main->V
MAC Address   = 00:20:a6:47:f7:30
IP Address    = 0.0.0.0
SSID          = socalfreenet.org
Channel       = 56
SNMP  Enabled = 0
AP or STN     = 0


Security Mode = 0
Default Key   = 1
WEP Key Size  = 13
Old wepState  = 0
Auth Address  = 0.0.0.0
Auth Address2 = 0.0.0.0
```

```
Auth Retry Tm = 0

Turbo Mode              = 0
Repeating Enbled        = 0
Beacon Interval         = 100
DTIM Period             = 1
Fragmentation Enabled   = 0
Fragmentation Threshold = 2346
RTS Threshold           = 2346
RTS Mode                = 0
Supported Rates         = 0xff
Turbo Supported Rates   = 0xff


keyBuf40 :  0 0 0 0 0 0 0 0
            0 0 0 0 0 0 0 0 0 0 0 0
keyBuf128:  0 0 0 0 0 0 0 0
            0 0 0 0 1 0 0 0 0 0 0 0 0 0 0 0
            0 0 0 0 0 0 0 0 0 0 0 0 0 0 0 0
            0 0 0 0 0 0 0 0 0 0 0 0
keyBuf152:  0 0 0 0 0 0 0 0
            0 0 0 0 0 0 0 0 0 0 0 0 0 0 0 0
            0 0 0 0 0 0 0 0 0 0 0 0 0 0 0 0
            0 0 0 0 0 0 0 0 0 0 0 0 0 0 0 0
            0 0 0 0 0 0 0 0
authSecret: 0 0 0 0 0 0 0 0
            0 0 0 0 0 0 0 0 0 0 0 0 0 0 0 0
            0 0 0 0 0 0 0 0 0 0 0 0 0 0 0 0
            0 0 0 0 0 0 0 0 0 0 0 0 0 0 0 0
            0 0 0 0 0 0 0 0

Main->
```

Another interesting menu can be found by pressing lowercase **f** and then the question mark (**?**):

```
Main->f
Radio Tests->?
a = set antenna
b = bc stats
```

```
c = set channel
d = dump eeprom
e = const dac
f = channel freq
g = pwr tx
h = pwr rx
i = init radio
j = stats
k = tx99
l = listen rx
m = tx loopback
p = set pwr ctrl dca
q = quit to main menu
r = set rate
t = set turbo mode
s = sine wave
x = continuous tx
y = continuous rx
Radio Tests->
```

Finally, another screen can be found by pressing lowercase **a** and then the question mark (**?**) to reveal the Atheros Radio menu:

```
Main->a
Radio->?
? = show help
a = display All error stats
A = set AP Mode
b = display station info
B = get MAC Reg
c = set channel
C = set MAC Reg
d = display config
D = DMA Size
e = rate Enable
E = display rate Counters
f = rate Disable
F = set Rate
g = set ch list
```

```
h = set turbo ch list
i = set hw tx retry count
I = set Beacon Interval
j = set RD display code
J = set DTIM Period
k = set repeating
K = display WEP Keys
l = radioCal
m = misc stat
M = display MAC regs
n = display Beacon
o = display semaphore
p = print radio stats
q = quit to main menu
r = reset radio
s = radio stop
S = radio Start
t = turbo mode
u = set RD
v = set anntenna type
V = set Turbo Allowed
w = set wep
x = dump EEPROM
X = dump Prox EEPROM
y = display Calibration
z = zero stats
0 = toggle Debug Flags
1 = set SIFS
2 = set DIFS
3 = set aggressive PIFS
4 = disable 48/96 and 54/108
5 = enable 48/96 and 54/108
6 = set Beacon txRate
7 = set BC MC txRate
8 = set EEPROM
9 = get EEPROM
Radio->
```

From this menu, you can modify all manner of wireless configuration options, including WEP keys, data rates, channels, regulatory domain (FCC, ETSI, Spain, France, and so on), and more. You can also display statistics and view a list of association stations.

NOTE

You should use extreme care in using the Debug mode and always remain in compliance with local regulations.

Summary

In this chapter, we reviewed firmware upgrades for the Linksys WRT54g AP as well as provided a review of the Soekris SBC hardware line. Finally, we reviewed the Proxim 8571 and how you can use it to create 802.11a links.

Choosing to use a Linksys or SBC device is a very deployment-specific issue. In general, we like to shy away from consumer-grade gear, but in some environments (such as small coffee shops or retail locations) it could be entirely appropriate.

Because upgrading Linksys firmware is so simple (just use the browser-based management interface), we recommend playing with multiple distributions before making your selection. For SBCs, always be sure to check the hardware requirements of your distribution before selecting a particular SBC product. Soekris engineering makes an excellent line of SBCs that work great in community wireless networks.

Another option to consider for backhauls is to use 5 GHz, where there is less interference and congestion than 2.4 GHz. A very low-cost method for building 802.11a backhaul links is to use a Proxim 8571. One device can operate as an AP while the other device can be "harvested" for its PCMCIA card and used as a client in a Soekris running pebble. Chapter 8 outlines other solutions that are commercial but low cost, such as the excellent Sputnik management platform.

## Chapter 5

Wireless Client Access Devices

Topics in this Chapter:

- Notebook Computers
- Desktop Computers
- Personal Digital Assistants
- WarDriving

Introduction

Let's say that you have just finished setting up your wireless network, or perhaps you want to connect to that free hotspot at your favorite coffee shop. To facilitate communication, a properly functioning wireless network requires an Access Point (AP) on one side and a wireless client access device on the other side.

So, what happens after you set up the APs? Well, you will need to be able to access that network somehow. In this chapter, we discuss the various types of client access to the wireless network. First, we show you how to connect using a notebook computer. Then, we show you how to hook up your desktop computer and Personal Digital Assistant (PDA).

By the time you're finished with this chapter, you will understand everything you need to know to get your client device up and running on a wireless network.

Notebook Computers

Notebook computers or laptops are by far the most widely used computing platform for accessing a wireless network. In fact, before the widespread use of wireless technologies became commonplace, most people had to either use a dial-up modem, or stretch a long, winding Ethernet cable around the room to connect to the Internet. However, you are now able to connect anytime and anywhere, regardless if it is from home, the coffee shop down the street, or sitting at the airport.

There are two main connectivity options for notebook computers; however, some of the desktop methods discussed later may work as well. The first connection device is a PCMCIA card in one of the laptop's card slots. The second is for some newer notebooks that have a mini-PCI slot.

PCMCIA Cards

PCMCIA cards (or "PC cards" as they are sometimes called) require a notebook with an available Type II card slot on the computer. The card contains both the 802.11 radio and antenna in a compact design. These cards used to be more expensive than their USB and PCI counterparts were, but due to the proliferation and critical mass of Wi-Fi, they can often be picked up for as low as $5 to $20 if you shop around.

There is not a great deal of variation between these cards, as they are fairly standard in design among the various vendors. The only real difference may be the chipset used for the 802.11 radio. The major manufacturers of wireless chipsets are Atheros, Broadcom, and TI (Texas Instruments).

Most Original Equipment Manufacturers (OEMs) only provide software and drivers for Windows and Mac operating systems in the packaging. However, if you search the Web, you can often find additional drivers for Linux, BSD, and UNIX.

Figure 5.1 A typical PCMCIA card (pictured Proxim Harmony 802.11a card)

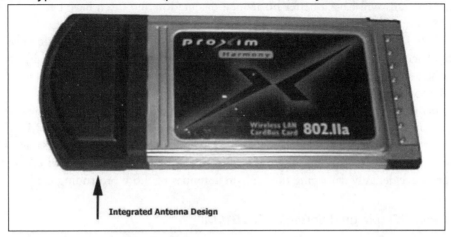

Integrated Antenna Design

As you can see in Figure 5.1, the antenna (the black part at the top of the card) extends out about half an inch or more from the card. This design is required to get better reception than if the antenna were buried inside the card slot.

The problem with this design is that the antenna is now vertically polarized and only receives the best signal both above and below the card. To compensate for this design flaw, some card manufacturers allow for the connection of an external antenna to increase performance as shown in Figure 5.2.

Figure 5.2 Another PCMCIA card (pictured EnGenius NL-2511CD PLUS EXT2)

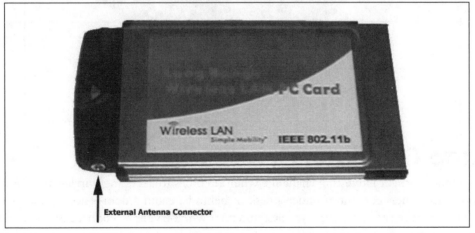

External Antenna Connector

Mini-PCI Cards

Mini-PCI cards are very similar to PCMCIA cards in design except that they lack the integrated antenna and preotective outer shell. These cards are designed for newer laptops that often have the antenna wiring built into the notebook behind the LCD screen. Because the antenna is behind the LCD screen, your cards will have a better horizontal orientation and often have better reception than their PCMCIA counterparts have.

Most mini-PCI slots are located on the bottom of the laptop under an accessible door similar to how one would access the memory or the hard disk. However, sometimes due to design constraints, we have seen manufacturers place mini-PCI slots under the keyboard, which requires a little more skill and finesse to access.

The antenna connectors of the card in Figure 5.3 are located in the upper left; they are the two little dots next to the large silver heat sink. Mini-PCI cards are more fragile than PCMCIA cards and are not designed to be removed and installed often. However, they are also very versatile, as you can upgrade your notebook's wireless card down the road and not have to worry about taking up a PCMCIA slot or accidentally damaging the built-in antennas of those protruding cards.

Figure 5.3 Mini-PCI (pictured EnGenius EL-2511MP)

Desktop Computers

Desktop computers are an interesting challenge when it comes to accessing wireless networks. Most people tend to have their computers under a desk or behind a cabinet door—not a good place to locate the PC (or more specifically, the wireless antenna) when trying to connect to a wireless network. Basically, in this situation you're placing a big piece of metal (the computer case) or large amounts of wood (the desk) between the radio and the AP. The signal will eventually penetrate, but with a loss of signal strength. In this section, we discuss your options when it comes to hooking up your desktop to an 802.11 network.

PCI Cards

PCI cards for desktop computers have come a long way in the past couple of years. Originally, they were implemented as a PCI-to-PCMCIA bridge that allowed you to insert a PCMCIA card into the back of your computer. The problem this created was that the antenna was again forced into a location that suffered from poor reception. Manufacturers then started to make PCMCIA cards with removable antennas to help alleviate some of this problem. Today, most PCI cards actually have the 802.11 radio built into the card instead of using a PCMCIA slot with the radio in a separate PCMCIA package.

As you can see in Figure 5.4, modern cards tend to have more powerful detachable antennas that can increase your reception. Some companies such as D-Link and SMC sell slightly more powerful omnidirectional and unidirectional antennas to increase performance and allow more flexibility in antenna placement.

Figure 5.4 PCI Card (pictured Linksys WMP11)

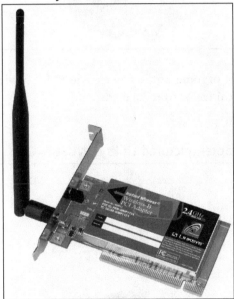

Courtesy of Lynksys.

USB Devices

USB radios offer some of the best flexibility for desktop computers. USB offers more deployment options than PCI because you can move the USB device around the room until you find its optimal orientation. Usually, they come with a six-foot USB cable, but if you are using a powered USB hub, you can go up to a distance of 15 feet from the PC. Shown in Figure 5.5 is an example of an ORiNOCO USB client adapter. These were very common just a few years ago. In fact the inside of the adapter is nothing more than a USB to PCMCIA bridge with a standard card sitting inside.

Figure 5.5 A typical USB adapter (pictured Lucent ORiNOCO USB Client)

Figure 5.6 is a great example of some of the newer style of USB wireless adapters. Most manufacturers have gone to a smaller form factor to reduce cost.

Figure 5.6 Another USB Adapter (pictured Linksys WUSB54G)

Courtesy of Lynksys.

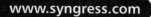

The only real downside to USB radios is the limited availability of drivers for the USB bus. Because of this problem, most USB devices only operate with Windows 2000 or XP. A few, however, are shipping with drivers for Mac OS.

Ethernet Bridges

Ethernet Bridges are wireless radios that can be used to extend a wireless network to an Ethernet switch or hub (which can be used to extend connectivity to multiple wired devices). Ethernet bridges can also be used to connect any device with an Ethernet port such as a Tivo, Xbox, or even a computer to the wireless network without having to install drivers or client software. This is a great solution for use with Mac OS and Linux computers, where drivers may be limited and more difficult to find.

Another benefit of using a wireless bridge is that since it uses wired Ethernet to deliver bandwidth to the client, you can extend the cat5 cable to its maximum segment length of 100 meters and still get connectivity. In theory, by using a Power over Ethernet (PoE) injector, you can send power over the Ethernet data cable as well and place the bridge as far away as 328 feet.

Most Ethernet bridges support external antennas. Figure 5.7 shows a Linksys WET11 with a removable RP-TNC antenna.

Figure 5.7 Ethernet Bridges (pictured Linksys WET11)

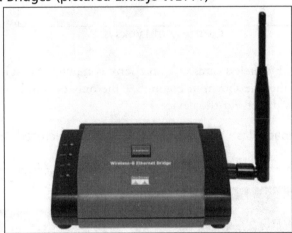

Courtesy of Lynksys.

PDAs

Personal Digital Assistants (PDAs) are growing in popularity. Just about anywhere you turn, someone has a Palm OS or Microsoft Pocket PC device. Wireless networking allows the ultimate in portable connectivity for handheld devices.

Compact Flash

Compact Flash (or CF) cards are the most common interface used by PDA devices. While originally used to extend the amount of memory in a device, the compact flash interface can now be used for network devices, such as the Linksys Compact Flash device shown in Figure 5.8.

Figure 5.8 Linksys Compact Flash 802.11b Network Interface

Courtesy of Lynksys.

You can even use most CF Wireless cards in a notebook computer through the use of a PCMCIA – CF adapter like the one shown in Figure 5.9, the only downside would if the card manufacturer never published any drivers for the device.

Figure 5.9 Another Compact Flash adapter w/ PCMCIA sled (pictured AmbiCom WL1100C-CF)

Secure Digital IO Cards

Relatively new to the market are the Secure Digital Input/Output (SDIO) cards. SD cards were also originally used to add memory and storage capacity to portable devices. Now, the SD interface has been extended to support network adapters (and is called SDIO). The advantage of these cards is that they are extremely small, lightweight, and require less power. Keep in mind that battery consumption is of paramount importance for PDA users. Some SDIO Wi-Fi cards also include storage space (memory) in addition to wireless functionality. Figure 5.10 shows a SanDisk SDIO Wi-Fi card with 256MB of built-in memory.

Figure 5.10 SanDisk SDIO Card with 256MB of Flash Memory

WarDriving

WarDriving is the act of discovering wireless networks. AP discovery can occur using a variety of transportation methods. In addition to driving around in cars, individuals have become very creative in their methods for seeking wireless LANs, capturing data by air, on foot, and by rail. What's next? Only time will tell.

First, let's review a little background on WarDriving. The term *WarDriving* has been credited to Pete Shipley, a security researcher from Berkley, California, who was one of the first people to automate the process of logging discovered wireless networks. Others had come before him, but they were manually logging APs with a notepad and pen. Because of his early pioneering work, Pete is often referred to as the father of WarDriving (more information about Pete can be found on his Web site at www.dis.org/shipley).

WarDriving can be accomplished using just about any notebook computer or PDA equipped with a Wi-Fi card. There is free software available for almost every operating system. Some of our favorites are the Linux distributions that are bootable from a CD-ROM and have all the tools you need preinstalled on the disk.

WARNING

There is a HOWTO document that is designed to function as a starting point for discovering wireless networks in an ethical and legal manner. It is never legal to access a secured AP through means of cracking encryption. The purpose of this HOWTO is for information gathering and data collection of historical trends. We are not responsible for any actions taken while WarDriving or how any information is used.

Why Are People WarDriving?

If you ask 10 different people who WarDrive why they do it, you will most likely get 10 different answers. When I started WarDriving, 802.11b was still fairly new, and I was curious as to how many people were using it either for personal or commercial use. I get excited when I see a pattern over time of more and more wireless networks in play. I also get discouraged when I don't see it being used in a secure manner, especially in businesses. I use my WarDriving data as a general audit of the region and use it as statistical data to encourage companies to lock down their wireless networks.

There are several organized WarDrives throughout the year. Some are in the form of contests such as the one held each summer at DefCon in Las Vegas, Nevada. Others are more of a collaborative effort, such as the WorldWideWarDrive (WWWD), which posts data about the entire United States and Europe on its Web site. WWWD is a massive worldwide coordinated effort to collect data during a one-week period.

NOTE...WWWD4 (JUNE 12–19, 2004)

During the WorldWideWarDrive in 2004, over 228,000 wireless networks were discovered and logged, which is an amazing number since the previous year only produced 88,000. During the week of the WarDrive, two SoCalFreeNet.org group members discovered over 19,000 APs in the San Diego region, 11,000 of which were unsecured.

More information about past and upcoming WarDrives can be found at www.worldwidewardrive.org.

Preparing for the Hack

For this hack, we have separated our materials into two lists: Required (including a notebook computer, wireless card, and some software), and Optional Equipment (including a global positioning system (GPS), power inverter, and external antenna). The Required list includes the basics you need to get started discovering wireless networks. The Optional list will enhance your experience and make your data more accurate and reliable.

Required Equipment

Getting started with the basics is easy—all you need is a computer and a wireless network card. Many free software applications are available to get you started.

Notebook Computer

For WarDriving, it is best to have an easily portable notebook computer. In theory, it is possible to use an old desktop computer with a wireless PCI card. However, this setup would take up a large amount of space in your vehicle, as it would require a monitor, keyboard, and mouse. With a tower PC, you can definitely rule out WarWalking (the act of discovering networks on foot versus using a car).

The operating system and discovery software that you use will determine the minimum system requirements needed. We personally would not recommend anything less than a Pentium III with about 256 MB of RAM, a 4–10 GB hard disk drive, and an available PCMCIA slot. If you plan to use a GPS to log your precise location where the wireless APs are discovered, you will also need an available serial or USB port. A suitable notebook can be found on eBay or other auction sites for less than $500.

Wireless Cards

When WarDriving, it is preferred to use a wireless card that supports an external antenna. Some brands popular among WarDrivers include Cisco, Orinoco Classic, and Senao. The latest versions of the NetStumbler and Kismet software now support a wider range of chipsets, including Hermes, Atheros, Atmel, Intersil Prism, and Cisco. For more information on wireless network cards and which chipset they use, refer to this excellent Web site: www.linux-wlan.org/docs/wlan_adapters.html.gz.

WarDriving Software

Many different software applications have been created to assist in WarDriving. In this section, we outline some of the more popular applications for each of the popular operating systems. Always be sure to check the requirements for each application and make sure you have the right kind of chipset (based on your particular client card) to support the software application you want to use.

Windows

An excellent and very popular wireless discovery tool designed for Windows is NetStumbler (www.netstumbler.com).

It has an extensive forum on its Web site dedicated to tweaking NetStumbler to meet your needs. As of this writing, the latest revision, v0.4, supports a wide array of cards to include Atheros, Prism, Atmel, and Cisco cards.

NetStumbler also supports scripting so that real-time mapping can be accomplished using Microsoft MapPoint or WiGLE.net's Java-based DiGLE mapping client (www.wigle.net).

Linux

Kismet (www.kismetwireless.net) is the most popular Linux wireless auditing application available. It has a wide range of tools built in, including WEP-cracking tools and real-time mapping of discovered APs. Kismet supports a wide range of cards to include 802.11a, 802.11b, and 802.11g chipsets.

Mac OS

For the Mac operating system, there are two popular choices: MacStumbler (www.macstumbler.com) and KisMac (http://kismac.binaervarianz.e). Both are free applications and can be downloaded from their respective Web sites.

MacStumbler requires an apple airport card and Mac OS 10.1 or greater. Unfortunately, there is currently no support for PCMCIA or USB devices. MacStumbler works by sending out probe requests with an SSID of "any" (as described in Chapter 1) and does not use monitor mode (which gives you the ability to analyze raw 802.11 frames). Therefore, MacStumbler will not detect closed networks.

KisMac is designed for OS X and does use monitor mode. No probe requests are sent and the application is totally passive. KisMac works with Orinoco, Prism, and Cisco cards.

BSD

dStumbler (www.dachb0den.com/projects/dstumbler.html) is an excellent AP discovery tool written by David Hulton of Dachb0den Labs. It operates in monitor mode and is totally passive. dStumbler includes support for Prism cards and has GPS support. dStumbler also has the capability (based on MAC address) to report if an SSID is at the default setting for that particular manufacturer.

Optional Equipment

The following items are optional and are not required to successfully WarDrive. They will, however, allow you record approximate locations of discovered wireless networks, extend the battery life of your notebook computer, and increase the number of APs discovered.

Global Positioning System

A GPS unit is essential for recording the latitude and longitude of discovered APs while WarDriving. Your WarDriving software will record this information when an AP is discovered. The accuracy depends on how fast you are driving and the effective range of the particular AP. It will most likely report accurately to within a few hundred feet or better.

There are GPS receivers available on the market for every budget and need. On the low end are GPS receivers that require a PC or PDA to process data and have no stand-alone mapping functions. These can generally be picked up at any local computer or office supply store for under $75.

Figure 5.11 Magellan SporTrack Handheld GPS Receiver

The next step up is handheld GPS units like the one shown in Figure 5.11. We prefer these units, as they are handy for more than just WarDriving. These GPS receivers can be used for hiking, mountain biking, or other outdoor activities, and almost all of them can track your route, by leaving "breadcrumbs" or dots on the screen. Some handheld GPSs will even show street-level details so you can see exactly where you are. This feature is extremely handy in metropolitan areas. A good handheld GPS will set you back about $150–$500 depending on its feature set and level of sophistication.

On the high end of the GPS product market are the vehicle-mounted units. These devices will often have full color displays and touch screens for entering driving or route information, and use DVD maps for street data. Due to their complexity and size, they usually require professional installation from your local car audio installer and will cost anywhere from $500 to over $1,000.

Regardless of which type of GPS you choose, you will need a way to get the longitude/latitude location data back into your computer. Most of the GPS units described have some type of data connection on the back of the unit that will allow you to connect via a serial or USB cable to your computer. On most of the handheld units, this port is shared with the external power connection, so look for a combo power/data cable if one is available. The format most commonly used for transmitting GPS data is known as the NMEA-0183 format.

On your notebook computer, you will need to configure a COM (serial) port to receive the data. Most often, if you are using a serial port it will be on COM 1. In addition, ensure that both your GPS unit and computer are set to the same speed—it is recommended that you use 9600, 8, N, 1, just like Figure 5.12.

Figure 5.12 Communications Port (COM1) Properties

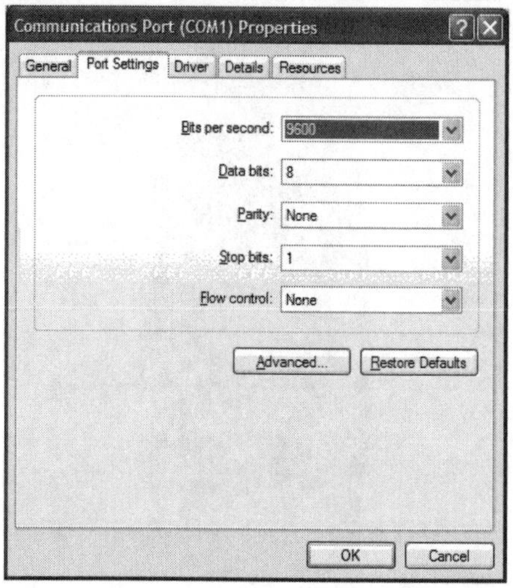

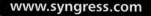

Power Inverter

If you're like most notebook owners, the batteries never seem to last long enough. Most notebook batteries will last an average of one to three hours depending on your power conservation settings. However, if you use a DC-to-AC power inverter, you can power your notebook during the entire WarDriving expedition and not be limited to the useful life of your batteries alone. Power inverters work by plugging into an available cigarette lighter jack, (similar to a cell phone charger) and often have one or more 120-volt (standard) AC receptacles on them. They are relatively inexpensive and can be picked up at most automotive supply or electronic stores for under $40.

Power inverters can be extremely handy, especially on long WarDrives. To adjust your power settings in Windows 2000 or XP, go to the **Control Panel** and select **Power Options**. When WarDriving with a power inverter, we recommend changing your power settings to match Figure 5.13.

Figure 5.13 Power Options Properties

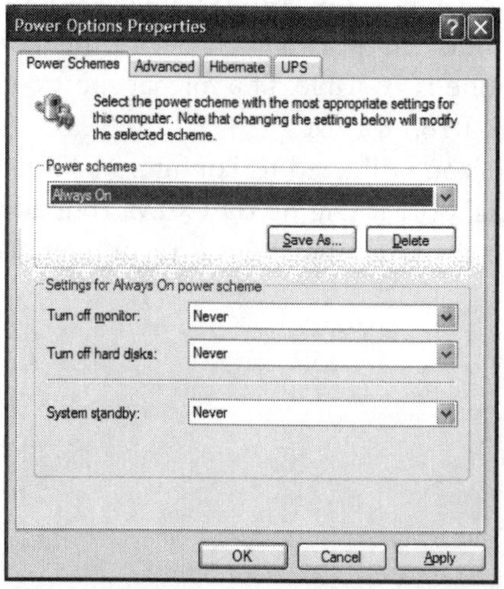

External Antennas

As mentioned earlier, it is highly recommended to use an external antenna while WarDriving. You will benefit from the higher RF gain and a more optimized signal polarization. This can mean the difference between discovering a few dozen or several hundred APs on your expedition. For more information on antenna types and design, refer to Chapter 10, "Antennas." Figure 5.14 shows what a small magnetic mount omni-directional antenna looks like.

Figure 5.14 Magnetic Mount Antenna

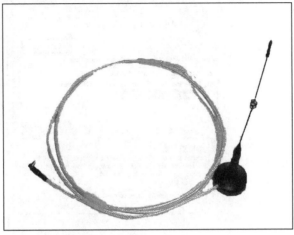

NOTE...THE LEGALITY OF WARDRIVING

Some people question whether WarDriving is legal or ethical. An excellent analogy can be made here with hiking. When hiking, you don't want to disturb the natural environment, so you pick up your trash and only take pictures instead of taking wildlife. Basically, you only leave footprints so others can enjoy the same beauty. As long as you take every possible measure to ensure that you are not actually connecting to these wireless APs and that all you are doing is the equivalent of taking a snapshot, then you can rest assured that your WarDriving activities are both legal and moral.

To ensure "safe" WarDriving, we highly recommend disabling your TCP/IP stack. Since wireless data networking occurs at Layers 1 and 2 in the OSI model, you can disconnect anything Layer 3 and higher and still engage in WarDriving activities. By disabling your TCP/IP stack, you are ensuring that your computer cannot have an IP address. Without an IP address, it is impossible for your computer to connect to any network or use any bandwidth or other resources.

To disable your TCP/IP stack in Windows:

1. Go to your **Network Connections** folder in the **Control Panel** and right-click on the applicable network adapter.

2. Select **Properties**.

3. In the **Components** section, uncheck the box next to **Internet Protocol (TCP/IP)** as shown in Figure 5.15. This will prevent you from being able to be assigned an IP address but still be able to actively scan for APs.

Figure 5.15 Network Connection Properties

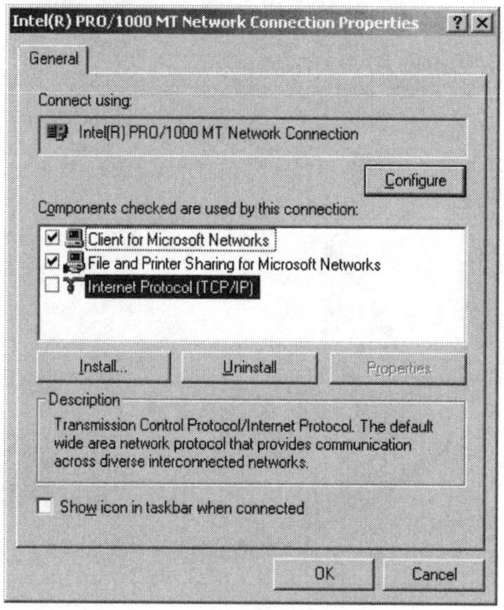

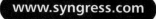

For more information on unbinding your TCP/IP stack in other operating systems, visit www.worldwidewardrive.org/nodhcp.html.

WarDriving Ethics

Finally, please keep in mind these suggestions for WarDriving:

- Do not connect to any networks unless you have explicit permission.
- Obey traffic laws.
- Obey property and no-trespassing signs.
- Don't use your data for personal (or monetary) gain.
- Set a good example for the WarDriving community.

What Can Be Done to Stop It?

If you do not want to have your network discovered by WarDrivers, there are precautions you can take to secure your wireless network.

- If your AP supports a "Closed Mode" feature, you can use it to disable the broadcast of the SSID in the management beacon. This is the actual data that most WarDriving packages look for. Note that this will not stop all WarDrivers, as some software can capture the SSID in the probe requests/responses from legitimate users who associate with the network.

- Turn on WEP or WPA security. This will not stop a WarDriver from locating your network, but it will prevent the casual WarDriver from automatically connecting to your network. In addition, be sure to change your keys often.

Other Resources

The following is a list of resources for finding out more about WarDriving:

- **Audit [of] Michigan Wireless** www.michiganwireless.org/staff/audit/wardriving

- **Stumbler Code of Ethics Renderman [of] Renderlabs** www.renderlab.net/projects/wardrive/ethics.html

- **The GPS Store** http://thegpsstore.com/pcpda_products.asp

- **WarDriving.com** www.wardriving.com

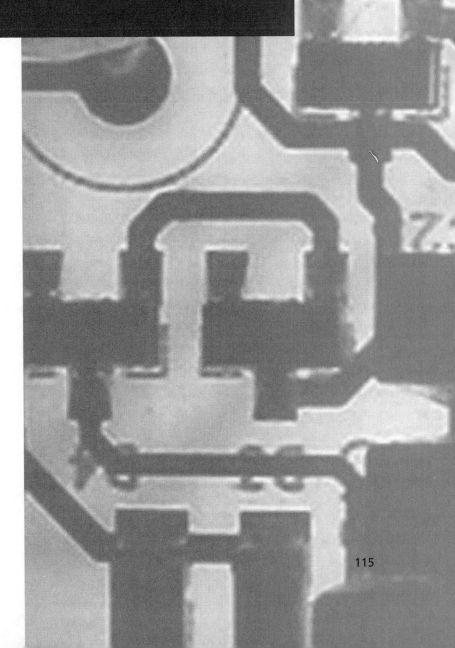

Part III

Software Projects

115

## Chapter 6

Wireless Operating Systems

Topics in this Chapter:

- m0n0wall—Powerful, Elegant, Simple
- Pebble—Powerful, Raw, Complete

Introduction

In Chapter 4, "802.11 Access Points" we looked at some of the hardware available to build your own access point. In this chapter, we'll examine some of the operating system distributions, or *distros,* you can use with the hardware. These are built upon either Linux or one of the BSD Unix derivatives like FreeBSD. In Chapter 8, "Low-Cost Commercial Options", we'll look at some of the commercially available operating systems that provide similar features for the same hardware platforms.

Every modern access point has an HTML interface to provide simple configuration via a Web browser. The operating system itself is hidden behind this customized user-friendly interface. In this chapter, you'll get below that interface and learn what's going on beneath the surface by installing your own distro. You can make your access point look as slick as a store-bought one, or keep it raw and geeky with the full power of the operating system at your fingertips.

Distros come in all shapes and sizes. Some provide nice graphical user interfaces, while others send you into raw Unix-like shell command prompts. Some are customized for old PC-compatible hardware, while others work best on newer single board computers (SBCs). Some assume you'll have a separate access point connected via an Ethernet port, while others can use a radio plugged directly into the computer. You'll probably try several solutions until you find the one best for you. Criteria to consider as you evaluate the alternatives are:

- What hardware are you using? An old PC, or a single board computer?

- Do you have a wireless card in the computer or a separate access point?

- What monitoring features do you need?

- How easy is it to configure?

- Can you easily make a backup of all settings?

Each of the distros in this chapter has key features that make it especially suited for specific tasks. Enthusiasts around the world have assembled a variety of distros and made them available on the Internet. We've chosen some of the more popular and powerful distros to show the broad range of features available, and to ensure you'll find upgrades and support. Table 6.1 summarizes some of the key features of each distro chosen:

Table 6.1 Operating System Distro Key Features

Distro	Hardware	Strengths	Comments
m0n0wall	SBC or PC with CD-ROM	Browser-based config-uration, easy installation and upgrade	A good starting distro
Pebble	SBC or PC with hard drive	Support for Atheros 802.11a, Debian Linux in 64MB	No browser interface—command line only

Installation techniques vary widely across distros and your chosen hardware. You'll need a second computer, apart from your access point hardware, to prepare the distro for use. Depending on the final medium, you'll also need either a compact flash (CF) writer or CD-ROM burner in your computer. Sometimes you can write to the CD or CF from a Windows or Macintosh machine, but often you'll need a machine running a variant of Linux or BSD. See Table 6.2 for a summary.

Table 6.2 Typical Hardware Install Media

Access Point Hardware	Typical Installation Media
Single board computer	Compact flash card
Old PC, no hard disk	CD-ROM drive, floppy disk drive
Old PC with hard disk	CD-ROM drive, for installation only

WARNING: HARDWARE HARM

Some of the utilities used for writing compact flash media are extremely cryptic. Further, since compact flash often appears to be just like an extra hard disk, it is relatively simple to accidentally overwrite on top of an existing hard disk, or other disk-like device such as a universal serial bus (USB) memory stick, instead of the compact flash card. Take extreme care to make sure that you are selecting the proper device when using compact flash utilities.

Some distros include support for radios installed on the hardware. This is especially true in the case of distros designed for single board computers, as these are usually installed outdoors on a mast (close to the antenna). Other distros provide features for computers connected via Ethernet to an external access point. They may provide special services for the access point such as captive portal, as well as more standard functionality like Dynamic Host Configuration Protocol (DHCP), Network Address Translation (NAT), and SNMP monitoring. Table 6.3 summarizes the hardware support for each distro.

Table 6.3 Type of Access Point Hardware Supported

Distro	Access Point Support
m0n0wall	Prism chipset or separate AP
Pebble	Prism, Orinoco, or Atheros chipsets, or separate AP

m0n0wall—Powerful, Elegant, Simple

The m0n0wall distro is the creation of a talented Swiss developer, Manuel Kasper. The home page is located at: http://m0n0.ch/wall , and uses zeroes, not the letter "O" in its spelling.

The m0n0wall distro stands apart from many others in several key ways:

- All configuration is performed via a Web browser.

- The configuration is stored in a single XML text file that can be easily saved and restored via the Web interface.

- It can be upgraded to a new version via the Web interface (if you use writable media such as compact flash cards).

The web configuration earns the most praise, which is appropriate. Manuel has an unusually good eye for design and his page layout is clean and uncluttered. Further, the usability is good, though the sheer technical complexity of the features available can be overwhelming.

Less noticed, but equally important, however, is the ease of backup and restore capabilities. There are few free distros that offer such a simple and efficient mechanism for quickly backing up *all* of the settings. Using an XML format is an added bonus, too, as it's easy to clone and tweak an existing configuration for installation on another machine. For example, just change the wide area network (WAN) IP address and install it on another device.

The following are other important features of m0n0wall when used as an access point, or when connected to an external access point:

- Network Address Translation

- Dynamic Host Configuration Protocol

- Caching DNS forwarder (simplifies client configuration)

- SNMP agent (for remote usage monitoring)

- Traffic shaper (limit the ability for one user to consume all of the bandwidth)

- Appropriate ISP support (DHCP client, PPPoE, PPTP)

- Stateful packet filtering

- Wireless AP support (with PRISM-II/2.5 cards)

m0n0wall has many more advanced routing and firewall features that are beyond the scope of this chapter. See the m0n0wall Web site (http://m0n0.ch/wall) for more information.

Preparing for the Hack

m0n0wall was originally written to run on single board computers as an open source equivalent of commercial Small Office/Home Office (SOHO) firewalls made by WatchGuard, SonicWALL, ZyXEL, and Netscreen. However, it was soon modified so that it would also work on most standard PCs. We'll cover both installation options in this chapter since the setup is the same once the appropriate media is prepared.

One interesting hybrid installation is to use a standard PC with a Compact flash to a hard disk Integrated Device Electronics (IDE) adapter. These adapters are typically around $20 and allow you to use a CF card just like a small hard disk. With this adapter, your old PC can be a firewall and/or access point and have no moving parts—which greatly increases reliability—assuming it is old enough not to require a CPU fan. The greatly reduced power load may also let you disable the power supply fan for completely silent operation.

m0n0wall on a Standard PC

To run m0n0wall on a standard PC, you'll need:

- A standard PC, 486 or better with 32MB of RAM (64MB recommended)
- Ethernet port and a PRISM II/2.5 wireless card, or two ethernet ports
- A CD-ROM and floppy drive, or, a small hard disk that can be completely erased
- CD-ROM writing software that writes ISO images (not supplied in Windows XP). Several freeware programs exist for XP, such as http://isorecorder.alexfeinman.com/isorecorder.htm or www.cdburnerxp.se. Search for "free iso burn software" to find others. Most Linux or BSD-based distributions will already have CD-ROM burning support available (for example, burncd or cdrecord).
- Another PC running either FreeBSD or Windows with a CD-ROM writer installed (or the target hard disk installed as a second disk drive)
- A blank CD-RW or CD-R disk
- A blank formatted 1.44MB diskette

The ethernet cards should be supported by FreeBSD/i386 version 4.9, as listed at: www.freebsd.org/releases/4.9R/hardware-i386.html. Most commonly available cards are supported.

m0n0wall on a Single Board Computer (SBC)

To run m0n0wall on a single board computer, you'll need:

- A single board computer made by Soekris or PC Engines, or a small form-factor standard PC with a compact flash "drive" (see Chapter 4 for more details on these boards).
- PRISM II/2.5 chipset wireless card (either PCMCIA or miniPCI, or PCI depending on your hardware).

- A compact flash card at least 8MB in size (larger is okay, but the extra space is not used and is unnecessary). Old digital camera cards are a great cheap source for the perfect-sized CF cards since they were usually quite small.

- Another PC running either FreeBSD, Linux, or Windows, with a compact flash writer installed. A suitable Windows CF writer is available at http://m0n0.ch/wall/physdiskwrite.php.

- (Optional, but recommended) Serial port on your main computer. In the event of problems, this is necessary for debugging and can simplify the initial configuration. Some newer computers don't have serial ports, so you may need a USB to serial port converter.

- (If using a serial port) A cable to connect the SBC console to your main computer's serial port.

- (If using a serial port) A serial terminal program such as Windows' included Hyperterm or the excellent Terra Term Pro from http://hp.vector.co.jp/authors/VA002416/teraterm.html.

Other Configuration Settings

To complete the m0n0wall configuration, you'll need your ISP (or WAN) settings, as shown in Table 6.4.

Table 6.4 m0n0wall ISP (WAN) Configuration Settings

Link Type	Settings Needed
Static IP	IP address, subnet mask, gateway address, DNS servers
DHCP assignment	DNS server addresses
PPPoE	Username, password, DNS server addresses
PPTP	Username, password, local IP address, remote IP address

ISPs may have various supplemental requirements to complete authentication, such as hostname for DHCP or a specific Media Access Control (MAC) address. m0n0wall has the ability to set most of these also.

If you have an existing LAN, you may also choose a different LAN subnet than the default supplied by m0n0wall (192.168.1.1/24), as well as many other configuration settings.

Performing the Hack

Now that you've picked up your hardware and obtained all of the configuration settings you'll need, it's time to get your hands dirty and start building. In this section, we will download the software and walk through the steps necessary to install the OS on your physical media and launch m0n0wall!

Downloading a Recent Version

The latest stable version of m0n0wall can be found at http://m0n0.ch/wall/downloads.php.

This will have the latest "stable" release, but it may be quite old. More recent versions can be found at http://m0n0.ch/wall/downloads/.

Each m0n0wall release tends to be quite stable, so don't be too concerned about going with something more recent, especially if it's been out for a few days already. You can check the e-mail list archive at http://m0n0.ch/wall/list/ to see if people are having problems with it.

Once you've found the files, you'll need to identify which version you need. Look in Table 6.5 for the hardware you plan to use:

Table 6.5 Summary of m0n0wall Download Versions

Hardware	Image Name
Compact flash image for Soekris 4501, 4511, 4521, and 4526 series of single board computers	Net45xx
Compact flash image for Soekris 4800 single board computer	Net48xx
Compact flash image for PC Engines WRAP board	Wrap
Compact flash or hard disk image for a standard PC	Generic-pc
CD-ROM ISO image for creating a bootable CD-ROM	Cdrom
FreeBSD file system image—handy for exploring and making changes to the m0n0wall distribution	Rootfs

Decide which version best suits your hardware and download it to your main computer. Then read the appropriate section that follows to find out how to install the image onto your computer.

Creating a CD-ROM from Windows

To create a CD-ROM, you'll need the file called cdrom-x.xx.iso (where the x's identify the version number). Then use your standard CD-ROM burning software to burn this onto a CD. Because an ISO file is an exact image of what was placed on the CD-ROM, this should be a simple process with your preferred CD-ROM burner software. The following images show how to do this with the ISO Recorder Power Toy from http://isorecorder.alexfeinman.com/ using Windows XP, but the basic steps are the same for any software:

1. Select the iso file to burn.

2. Select the CD-ROM burner drive.

3. Write it!

Step by Step Using the ISO Recorder Power Toy

Perform the following:

1. Locate the file you downloaded. It will have a name similar to cdrom-1.1b9.iso.

2. Place a blank CD in your CD burner, and then right-click the file and choose **Copy Image To CD** from the menu, as shown in Figure 6.1.

Figure 6.1 Create a CD from an ISO Image Using the ISO Recorder Power Toy

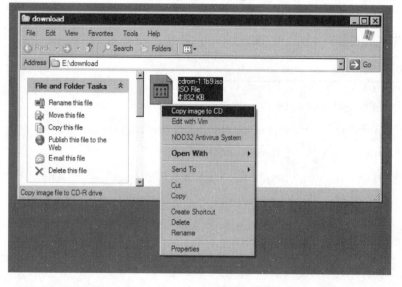

This will start the wizard. It verifies the source file and the destination CD burner, as shown in Figure 6.2.

Figure 6.2 ISO Recording Wizard Confirmation

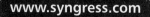

If you don't have a blank CD in the burner, it won't let you continue and you'll need to cancel the wizard and start over. The Recorder Properties should be set correctly, but if you have trouble making a usable CD, you can use the settings to write the CD more slowly.

The CD writing process should be fast, even for a slow CD writer, as less than 10MB are written to the 650MB capacity of the CD. Once the CD is finished, you're ready to boot m0n0wall!

Creating a Compact Flash (CF) Card from Windows

To create a compact flash card version of m0n0wall, you'll need the appropriate image file for your target machine, as shown in Table 6.5.

WARNING: HARDWARE HARM

It's vital to verify which device corresponds to your CF card, because you can easily overwrite your (primary!) hard disk or other storage devices like USB memory keys with this utility. This is discussed in detail in the next section, so pay attention and don't skip ahead! We suggest you remove any nonessential storage devices before attempting to write to your CF card.

The steps to safely write m0n0wall to your CF card are:

1. Download the appropriate image file.
2. Remove the CF card from your reader if it's already inserted.
3. Run the physdiskwrite program.
4. Note the drives available.
5. Cancel the physdiskwrite.
6. Insert the CF card again.
7. Run the physdiskwrite program again.
8. Compare the drives available and confirm that the new drive appears to match the size and other details of your CF card.
9. Confirm the copy to the CF card.

You should repeat steps 2 through 8 until you are certain your card is being recognized and that you know which device it is.

The following detailed example will assume you're using a PC Engines WRAP board, but the strategy is identical for all CF-powered versions. Locate the file you downloaded. It should have a name like wrap-1.1b9.img. Remove the CF card if it's already inserted. Open a command prompt using **Start | Run** and enter **cmd** (or **command** if you're running Windows 98). Now you'll run physdiskwrite using **Physdiskwrite wrap-1.1b9.img**.

Immediately press **Ctrl + C** on the keyboard. This will generate output like that shown in Figure 6.3.

Figure 6.3 Results of physdiskwrite with CF Card Removed

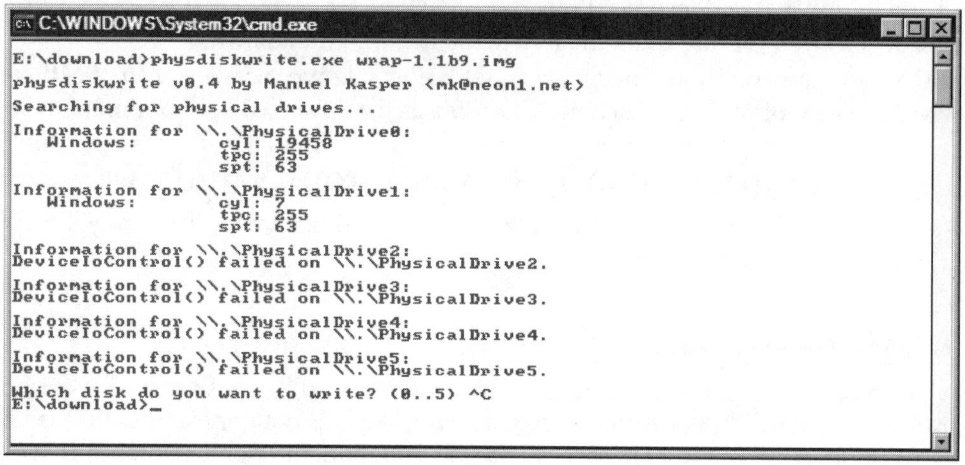

```
C:\WINDOWS\System32\cmd.exe

E:\download>physdiskwrite.exe wrap-1.1b9.img
physdiskwrite v0.4 by Manuel Kasper <mk@neon1.net>
Searching for physical drives...
Information for \\.\PhysicalDrive0:
    Windows:          cyl: 19458
                      tpc: 255
                      spt: 63
Information for \\.\PhysicalDrive1:
    Windows:          cyl: 7
                      tpc: 255
                      spt: 63
Information for \\.\PhysicalDrive2:
DeviceIoControl() failed on \\.\PhysicalDrive2.
Information for \\.\PhysicalDrive3:
DeviceIoControl() failed on \\.\PhysicalDrive3.
Information for \\.\PhysicalDrive4:
DeviceIoControl() failed on \\.\PhysicalDrive4.
Information for \\.\PhysicalDrive5:
DeviceIoControl() failed on \\.\PhysicalDrive5.
Which disk do you want to write? (0..5) ^C
E:\download>_
```

Some of the details to notice in this output are:

- Two physical drives present, one of which is quite small. This is a USB memory key that is inconvenient to remove. The other is the main hard disk for the system!

- Four physical drives which return an error. These correspond to a multi-format card reader with no cards in it.

Once you understand which disks correspond with which, you're ready to insert the CF card and run the same command again. Now the output will change to something similar to what's shown in Figure 6.4.

Figure 6.4 Results of physdiskwrite with CF Card Inserted

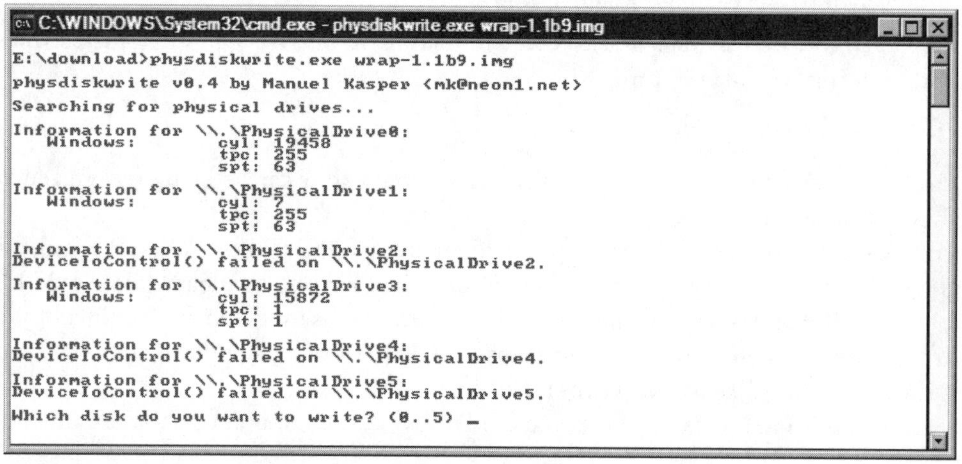

```
C:\WINDOWS\System32\cmd.exe - physdiskwrite.exe wrap-1.1b9.img

E:\download>physdiskwrite.exe wrap-1.1b9.img
physdiskwrite v0.4 by Manuel Kasper <mk@neon1.net>
Searching for physical drives...
Information for \\.\PhysicalDrive0:
    Windows:          cyl: 19458
                      tpc: 255
                      spt: 63
Information for \\.\PhysicalDrive1:
    Windows:          cyl: 7
                      tpc: 255
                      spt: 63
Information for \\.\PhysicalDrive2:
DeviceIoControl() failed on \\.\PhysicalDrive2.
Information for \\.\PhysicalDrive3:
    Windows:          cyl: 15872
                      tpc: 1
                      spt: 1
Information for \\.\PhysicalDrive4:
DeviceIoControl() failed on \\.\PhysicalDrive4.
Information for \\.\PhysicalDrive5:
DeviceIoControl() failed on \\.\PhysicalDrive5.
Which disk do you want to write? (0..5) _
```

Now there is a "PhysicalDrive3" that wasn't there before. To double-check, the numbers should all be smaller than "PhysicalDrive0," which is the main hard disk for the computer.

You should repeat the physdiskwrite command several times with and without the card inserted until you're absolutely sure you'll be writing to the correct disk. When you're certain, you can enter the number (**3** in this example), and you'll get a confirmation prompt. Press the **Y** key to continue, or **N** to cancel, followed by the **Enter** key. The data will then be written to the CF card and a counter showing the progress will be displayed. When writing is complete, a confirmation message will appear, as shown in Figure 6.5.

Figure 6.5 Completed Output of physdiskwrite

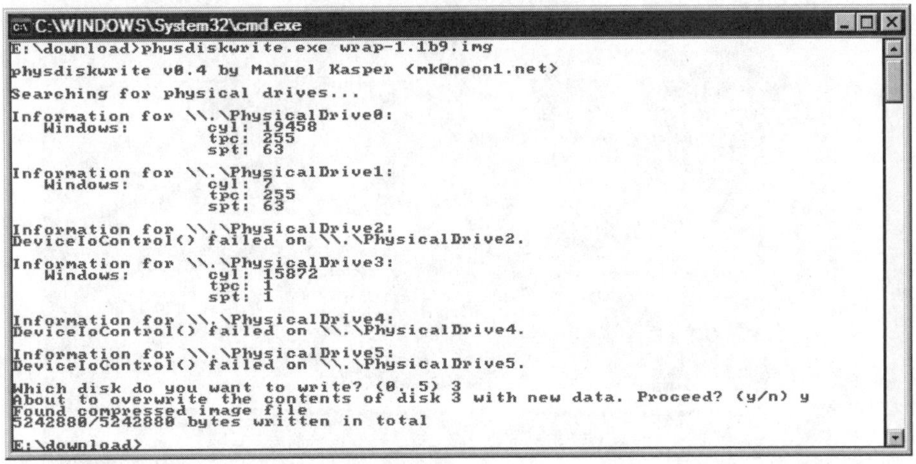

If the write completes successfully, you're now ready to put the CF card into your other computer and turn it on!

Starting Your Standard PC

Now you have all the pieces together to start your standard PC. This section takes you step by step through the process of turning your old PC doorstop into a modern firewall and access point.

Booting from the CD-ROM and a Blank Diskette

If you're using a CD-ROM and diskette, be sure you first change the boot order for your computer. You can make this change in your BIOS settings (described in a moment). It is important that the blank floppy is available when you first boot m0n0wall from CD because it only checks for its existence at boot time and will only create an empty configuration file at boot time. It is tempting to not change the boot order and to try and insert the floppy at "just the right time" after the CD has begun booting, but we found that this is harder than just changing the boot order in the first place.

The boot order configuration is set in the BIOS of your computer and can be changed when it first starts up by pressing a specific key such as **F2** or the **Delete** key—it will usually tell you as it boots. Then find the appropriate setup screen for setting the boot order. Figure 6.6 shows a typical configuration screen with CD-ROM Device at the top of the list, Hard-Disk Drive C: at the bottom, and the 3.5" Diskette second. This means, of course, that if you don't insert a floppy or CD, the computer will boot normally from the hard disk. This is convenient for testing and configuration since you can still boot from your hard disk if need be. This is handy if you're testing m0n0wall on a different PC from the final machine you'll use, or if you just need a firewall temporarily.

Figure 6.6 A Typical Boot Order Configuration Screen

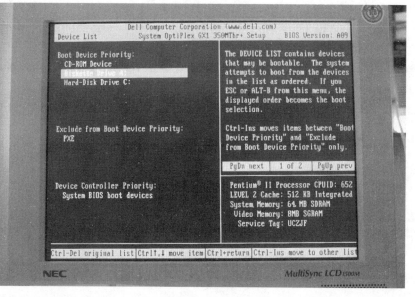

NEED TO KNOW...DEFAULT INTERFACE ASSIGNMENT

By default, m0n0wall will make the "first" Ethernet port the LAN port and use the second port for WAN. If you can identify which is which, you can skip the console configuration steps described in this section. For single board computers, the LAN port will be the eth0 port (see Table 6.6). For standard PCs, you can try first one port and then the other to see if you get an IP address via DHCP. This will typically be 192.168.1.199. If you have a wireless card installed, it will not be automatically enabled or assigned by m0n0wall. However, you can do that from the web interface once you've logged in.

Table 6.6 Single Board Computer Configuration Information

Product	Default Serial Speed	Interrupt Boot Key	Eth0 Port when facing ethernet connectors
Soekris	19200	Control – P	Right Hand Side
PC Engines WRAP	38400	S	Right Hand Side

Assigning m0n0wall Network Interfaces

The m0n0wall console allows you to configure your network ports. If you're using an older 10Mbps-only network card, you may wish to assign that to your broadband DSL or Cable connection since it's unlikely to exceed 6Mbps, and then use your other 100Mbps card for the LAN connection. Figure 6.7 shows the console menu for m0n0wall. There are several options available, but the only thing you need to do with the console is to map your network cards to their function—for instance, WAN or LAN. If you have more network cards, you can either assign them here, or do so later using a Web browser. For security, you can disable the console option completely once you've logged in via a Web browser.

Figure 6.7 The m0n0wall Console Setup Screen

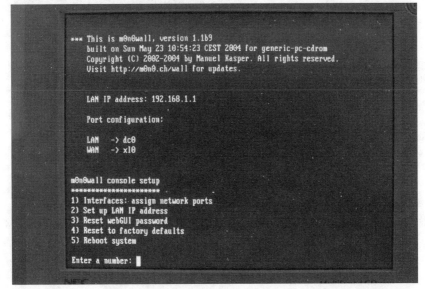

Choose option 1, "Interfaces: assign network ports" by pressing **1** and then the **Enter** key, as shown in Figure 6.8. If your network cards are recognized successfully, you'll see them listed under the heading: "Valid interfaces are." If you have them connected to an active device, their MAC address will be followed by "(up)".

Figure 6.8 m0n0wall Assign Network Ports

```
        LAN   -> sis0
        WAN   -> sis1


m0n0wall console setup
***********************
1) Interfaces: assign network ports
2) Set up LAN IP address
3) Reset webGUI password
4) Reset to factory defaults
5) Reboot system

Enter a number: 1

Valid interfaces are:

dc0      00:04:5a:74:af:15
xl0      00:c0:4f:5b:b9:31

If you don't know the names of your interfaces, you may choose to use
auto-detection. In that case, disconnect all interfaces before you begin,
and reconnect each one when prompted to do so.

Enter the LAN interface name or 'a' for auto-detection: █
```

m0n0wall includes a convenient auto-detection mechanism that works by following these steps:

1. Unplug all cables from the ethernet cards in your standard PC.

2. Type **A** for auto-detection.

3. Plug in the ethernet cable for the interface it requests (LAN, WAN, or something else).

4. Repeat the steps for each interface.

Assuming your cables are wired correctly and the devices they're connected to are running correctly, m0n0wall will detect that you plugged in the cable and then automatically assign that network card to that function. Figure 6.9 shows the results.

Figure 6.9 m0n0wall Network Port Assignment Completed

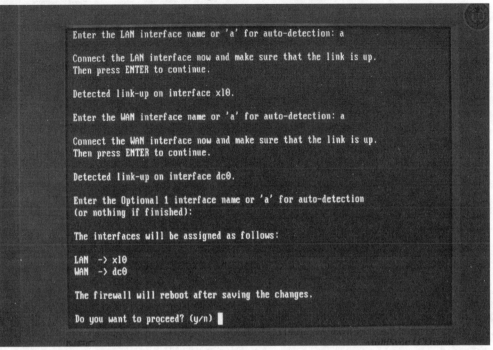

```
Enter the LAN interface name or 'a' for auto-detection: a

Connect the LAN interface now and make sure that the link is up.
Then press ENTER to continue.

Detected link-up on interface xl0.

Enter the WAN interface name or 'a' for auto-detection: a

Connect the WAN interface now and make sure that the link is up.
Then press ENTER to continue.

Detected link-up on interface dc0.

Enter the Optional 1 interface name or 'a' for auto-detection
(or nothing if finished):

The interfaces will be assigned as follows:

LAN  -> xl0
WAN  -> dc0

The firewall will reboot after saving the changes.

Do you want to proceed? (y/n) ▌
```

Once you've completed the network assignment, you can type **Y** and press **Enter** to save the data and reboot your firewall. Once it's restarted, you're ready to continue with the rest of the configuration using the browser.

Starting Your SBC

Installing m0n0wall on your single board computer (SBC) is similar to a standard PC, but you'll need to connect to your SBC via a serial port, rather than a keyboard and monitor, so you can access the console. You should also install any radio card you wish to use, though it's not necessary to connect the antenna at this point. Figure 6.10 shows the PC Engines WRAP.1C board all ready to configure. At the top left you can see the 8MB CF card with a new installation of the m0n0wall wrap distro, the serial cable is connected at the bottom left, and the radio card is in the left-hand miniPCI slot. Power is connected via the bottom right-hand connector.

Figure 6.10 A PC Engines WRAP.1C Board Ready to Configure

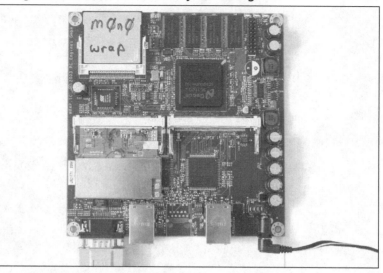

WARNING: HARDWARE HARM

Make sure the CF card is firmly in place. In the WRAP board, it's easy to catch the raised lip at the back of the card on the edge of the circuit board and not seat the card correctly. So be careful.

Now you'll need to run your terminal program and configure it for your SBC. The WRAP board by default uses a baud rate of 38400. You can leave all the settings except baud rate at their default values, which will usually be 8-bit data, no parity, 1 stop bit. Table 6.6 is a handy reference for boards mentioned in this book. In Tera Term Pro, use **Setup | Serial Port ...** to show the screen in Figure 6.11 and set the speed to 38400.

Figure 6.11 Tera Term Serial Port Setup

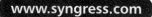

Now you're all ready. Apply power to your board. The exact sign on display screen will vary depending on the board, but if you've set the speed correctly and your serial cable wiring is correct, text will appear immediately after you apply power. When it does, immediately press the appropriate key to interrupt the boot sequence. Again, each board will be different. As you can see in Table 6.6, you press the **S** key for the WRAP board. You should then have output matching Figure 6.12.

Figure 6.12 A PC Engines WRAP Board Powerup Menu

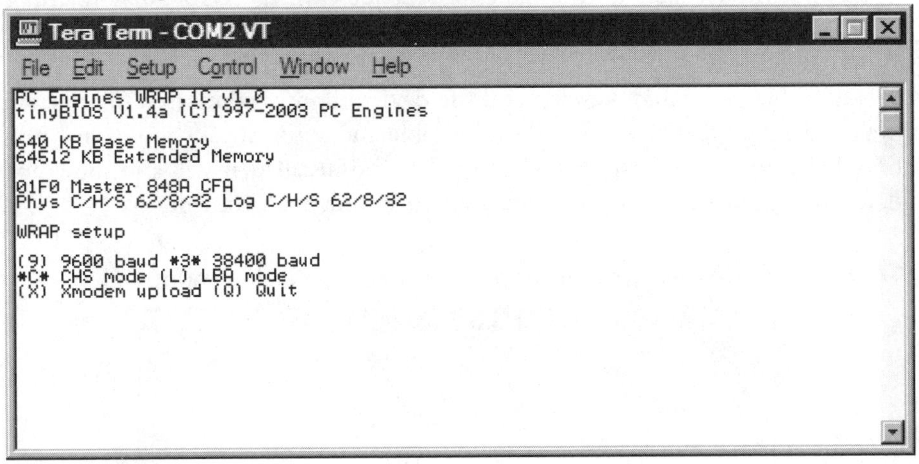

Now you need to set the default baud rate to 9600 to match what m0n0wall uses for the console. You do this by pressing **9**, **Q** to quit, and then **Y** to save the changes. There will be a short pause and then you'll see gibberish on the screen as the board reboots to a different speed.

Remove power from the board, change your serial port speed again, and then re-apply power. This time, don't interrupt the boot process and you should eventually see the display shown in Figure 6.13.

Figure 6.13 The m0n0wall Console Menu

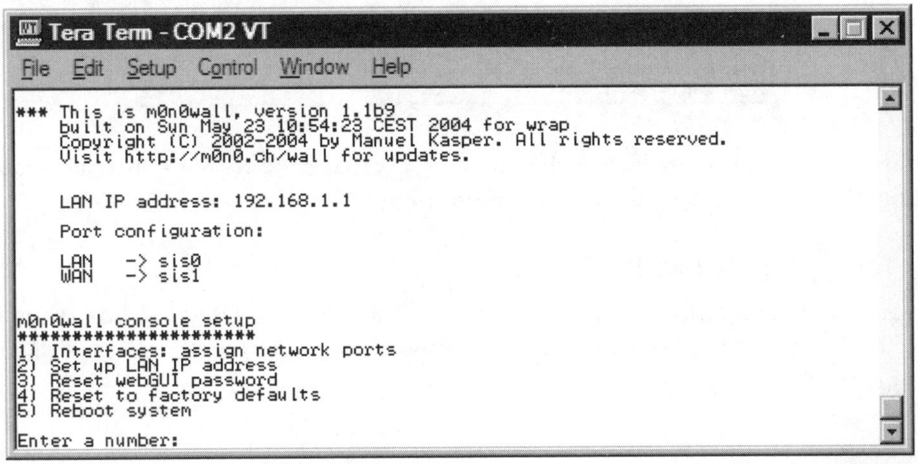

Once you see the m0n0wall console menu, you can assign the interfaces using the convenient auto interface feature of m0n0wall. This works by following these steps:

1. Unplug all cables from the ethernet cards in your standard PC.

2. Type **1** to assign interfaces.

3. Type **A** for auto-detection.

4. Plug in the ethernet cable for the interface it requests (LAN, WAN, or something else).

5. Repeat the steps for each interface.

Assuming your cables are wired properly and the devices they're connected to are running correctly, m0n0wall will detect that you plugged in the cable and automatically assign that network card to that function. Because we also have an 802.11b radio card installed, it will also show up, with the name wi0. Figure 6.14 shows the complete step-by-step interface assignment.

Figure 6.14 Assigning the Network Interfaces and Radio Card in m0n0wall

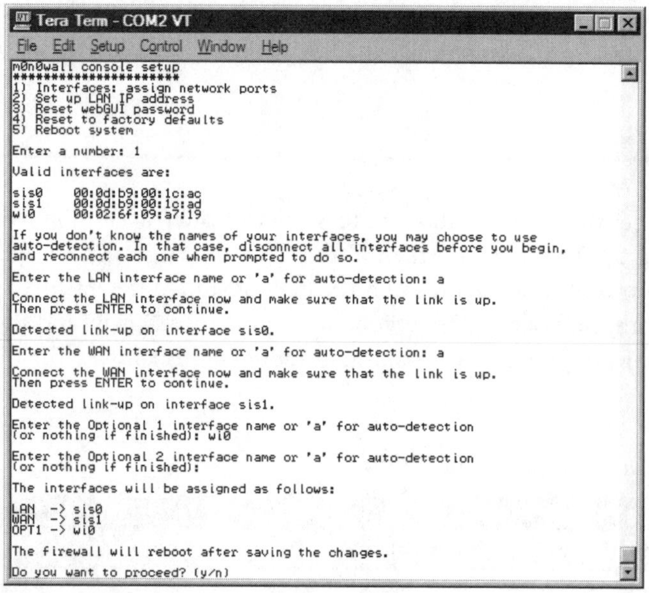

Once the interfaces are assigned, you should enter **Y** to save and reboot. Once it's restarted, you're ready to continue with the rest of the configuration using a Web browser.

Configuring m0n0wall

Now that you have your interfaces assigned, you're ready to log in to m0n0wall from your Web browser. For best results, you'll need a recent Web browser such as Internet Explorer or Mozilla using the information in Table 6.7.

Table 6.7 m0n0wall Web Browser Login Information

Item	Value
URL Address	http://192.168.1.1
Login Name	admin
Login Password	mono (all lowercase, letter o, not number 0)

Type in the URL http://192.168.1.1, which should result in a standard browser login prompt. The initial username is admin, and the default password is mono, which is all letters and all lowercase (no zeroes). Table 6.7 summarizes this information. Once you've completed the login, you should see the screen shown in Figure 6.15.

Figure 6.15 The m0n0wall Admin Interface

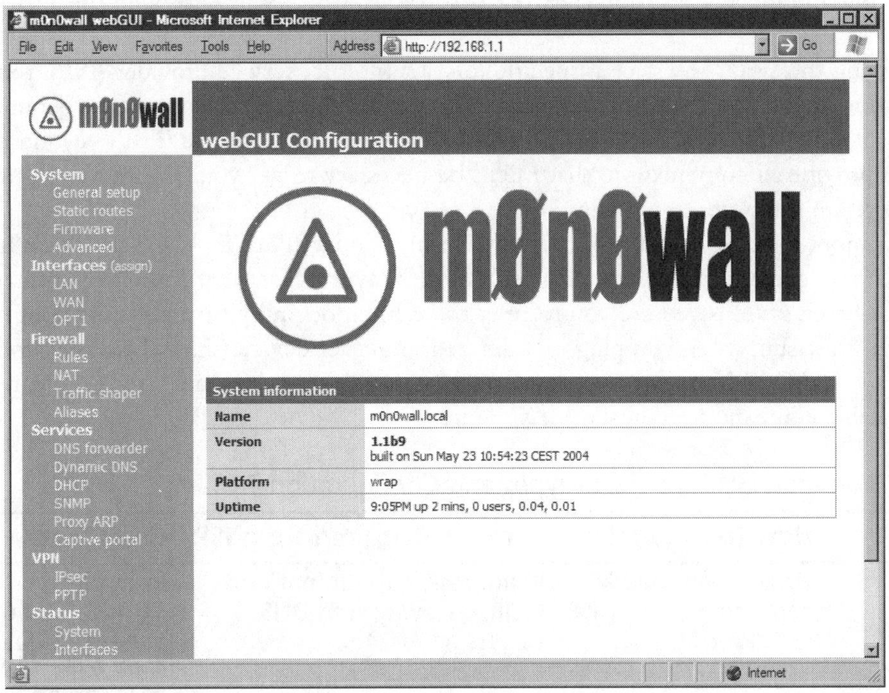

If you don't see this screen, chances are your computer is not on the right subnet. Make sure you've set your Internet connection settings to use DHCP and have verified that your computer's IP begins with 192.168.1.x. If it starts with something else, then you don't have your main computer's network settings configured correctly. Look for a setting called DHCP or another option titled "Obtain An IP Address Automatically."

This screen confirms the version and target machine (WRAP in this example) as well as how long the firewall has been running. You can get back to this screen if you click the **Status | System** link in the left-hand menu bar.

Before going much further, it's best to get the time set correctly and to change the administrator password. This is done by clicking **System | General Setup**, the first link in the left-hand menu bar. Here you should change the Password and then scroll down to the Time Zone drop-down. Select the closest city to your location. Leave the rest of the settings for now (even seemingly important ones like the DNS servers). We'll get back to them in a moment.

Be sure to click the **Save** button before continuing.

WAN—Get Online

If your main Internet connection happens to be DHCP-enabled, you may already be online. You can verify this by opening a Web browser and pointing it to your favorite Web site. If that works, you can skip this section and move on! If not, click the **Interfaces | WAN** link to open the WAN configuration page.

Configuring the WAN interface is specific to your Internet service provider (ISP). You'll need to refer to the documentation that they provided to be sure of the type of connection parameters you have. If you already have a computer online, then you can use that as a guide. It's beyond the scope of this book to provide all combinations that might be necessary to get your WAN port configured correctly, but here are some tips to get you going.

The commonest connection types are DHCP, Static, and PPPoE. Table 6.8 has some hints and guidelines for recognizing and configuring each of these types. Note that some ISP connections, notably cable-based systems, require you to reset the cable modem by turning it off, waiting a minute and turning it on again, when you plug in a different ethernet device (since the cable modem will be looking for the MAC address of the previously connected device). In some rare cases, you may need to call your cable company to reset the MAC address.

Table 6.8 Common ISP Connection Types and Configuration Tips

Type	How to Recognize	Settings Required from ISP	Comments
Static	Usually very clearly communicated by ISP as "static IP"	IP address, subnet mask or CIDR, gateway, and DNS servers	Example: *IP: 123.3.24.67* *Subnet: 255.255.255.0* *Gateway: 123.3.24.1* *DNS: 22.33.44.55,* *22.33.46.55* DNS values are entered on the **System	General Setup** page
DHCP	Usually nothing specified by ISP	None typically; may require a specific hostname be set	(none)

Continued

Table 6.8 Common ISP Connection Types and Configuration Tips

Type	How to Recognize	Settings Required from ISP	Comments
PPPoE	ISP usually provides a setup disk along with a username and password.	Username, password	If your computer currently has PPPoE in use, disable it before trying to reconnect via m0n0wall since it will no longer be necessary.

m0n0wall uses Classless Inter-Domain Routing (CIDR) addressing instead of the older subnet mask style. Common mappings are shown in Table 6.9. Use this as a guide for your static IP configuration, and other subnet settings in this chapter.

Table 6.9 Common Subnet Mask–to-CIDR Conversions

Subnet Mask	CIDR Equivalent
255.0.0.0	/8
255.255.0.0	/16
255.255.255.0	/24
255.255.255.224	/27
255.255.255.254	/31

After you enter the settings and click **Save**, you'll need to restart m0n0wall as prompted by the system, or by clicking **Diagnostics** to open the Diagnostics menu, and then choosing **Reboot System**. This will take a minute or two.

You should then test your WAN connection before continuing. Click **Diagnostics** and then **Ping**. Enter a hostname that you can normally reach and click the **Ping** button. The results should look similar to those in Figure 6.16.

Figure 6.16 Results of a Ping Test

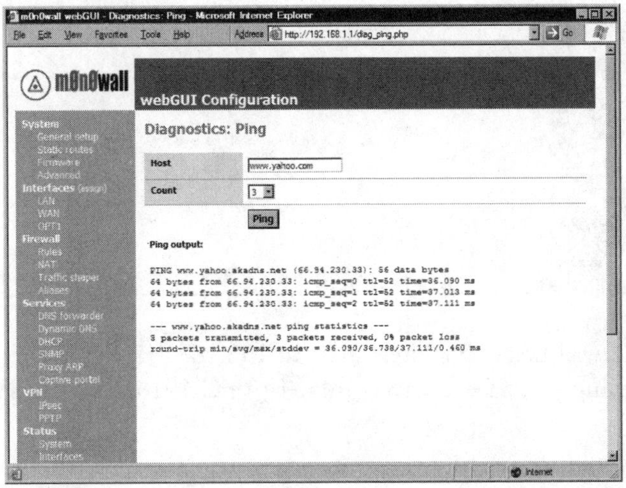

If the ping is successful, then your WAN link is working. If it fails, then next try to ping a known Internet address. For example, if you know your DNS or gateway address (provided by your ISP), you could try pinging them. For instance, instead of filling the **Host** field with a web site name like www.yahoo.com, as shown in Figure 6.16, you would use an IP address like 66.94.230.33. If an IP address works but a text address won't, then your DNS settings are incorrect or missing.

If ping fails for both text and numeric IP addresses, the next step is to check m0n0wall's logs under **Diagnostics | System logs**. The bottom of the log page contains the most interesting information, as the first 20 to 30 lines are system startup information and are not relevant for WAN configuration. The errors will vary by the type of ISP connection you have. Look for clues like *error* or *failed* to help determine what is failing.

If you continue to have problems, search your ISP's help pages or ask your ISP for assistance. It is becoming common for households to have more than one computer (or other Internet-connected devices). Most people use an Internet sharing device often called a home gateway or broadband router (or something similar), which all require the same settings as your m0n0wall device.

LAN—Customizing for Your Network

Once your m0n0wall is on the Internet, you should be able to immediately use it from the same computer you used to set up the m0n0wall configuration. By default, it will forward local traffic and route the responses back to your computer. The basic LAN configuration is set using the **Interfaces | LAN** screen as shown in Figure 6.17. The only option available is to set the LAN IP address of the m0n0wall and the network mask length.

Figure 6.17 The m0n0wall LAN Configuration

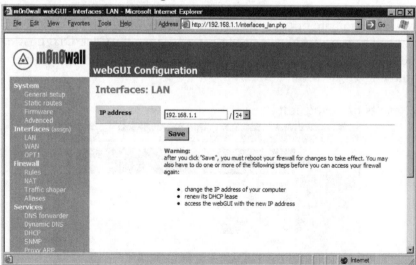

By default, the m0n0wall LAN IP address is set to 192.168.1.1, but you can choose any address or subnet you desire. Assuming you have a private subnet (RFC 1918), you can safely pick any address

range shown in Table 6.10. It's common for a gateway device like the m0n0wall to run at the ".1" address, so it's recommended you do the same.

Table 6.10 Reserved Private IP Address Ranges

Network Range	CIDR length	Comment
0.0.0.0 to 0.255.255.255	/8	A subnet of this is handy to type and remember—for example, 10.10.10.1/24.
72.16.0.0 to 72.31.255.255	/16	Less commonly used, so it may prove handy in avoiding confusion with other networks.
92.168.0.0 to 92.168.255.255	/24	The lower ranges are often used as default settings for vendor equipment—for example, many home LANs run on 192.168.0/24 or 192.168.1/24 subnets.

Once you make changes to the LAN IP, you'll need to reboot your m0n0wall using the **Diagnostics | Reboot system** menu, and then possibly restart all the computers on your LAN also (or force a release/renew of their IP addresses).

Once you have the LAN configured to your satisfaction, you can plug the m0n0wall LAN port into a hub or switch and multiple computers will be able to share the one Internet connection. Plug an access point directly into the m0n0wall or into the hub and you'll immediately have wireless access as well.

The external access point is transparent to the m0n0wall and the rest of the network—it appears to be another hub with one or more computers hooked up to it. Those wireless computers will still use the m0n0wall to lease an IP address and receive NAT services for shared Internet access as well. You'll need to configure the access point via its configuration interface so that its management IP address is on the same network as the m0n0wall. For example, if your m0n0wall is at 192.168.1.1, then you might assign your wireless access point the address 192.168.1.2 so it doesn't conflict.

You can either assign the AP an IP directly via its management interface (usually via a Web browser), or you can use the DHCP static assignment feature of m0n0wall to allow it to retrieve its setting via DHCP. To configure static DHCP addresses, go to the **Services | DHCP** menu. Scroll down to the very bottom and click the + (plus) symbol on the right-hand side. You can then enter the desired IP address (for instance, **192.168.1.2**). Then find the MAC address of the device, usually marked on the outside of the box, and enter that (say, **00:80:C8:AC:F8:64**). And finally enter a description such as **Upstairs 802.11b Access Point**. Click **Save** and you'll see the entry at the bottom of the DHCP table, as shown in Figure 6.18.

Figure 6.18 m0n0wall DHCP Lease Configuration

You can also enter MAC addresses from your computers into this table so they'll receive the same IP each time they request one. You can check the m0n0wall DHCP logs at **Diagnostics | DHCP leases** and then click the **DHCP** tab to see recent leases made to computers on your network.

Access Point—Turning on the Radio

If you have a wireless radio in your computer, now it's time to turn it on so wireless users can also access the Internet. (If you have an external access point you wish to use, simply plug it into the LAN port as described in the previous section.)

If you have the menu item **Interfaces | OPT1** then skip ahead. If you just installed the radio, then you'll need to add it via the small (**assign**) link to the right of the bold **Interfaces** item on the menu. Click the + (plus) sign below and to the right of the table of interfaces. If the wireless card is recognized and is the only new interface added, it will automatically create a new OPT1 entry and assign the radio card, as shown in Figure 6.19. The radio card will typically be called wi0, for "wireless zero," whereas Ethernet interfaces are called sis0 or eth0, depending on their chipset.

Figure 6.19 Newly Added OPT1 Radio wi0 Interface in m0n0wall

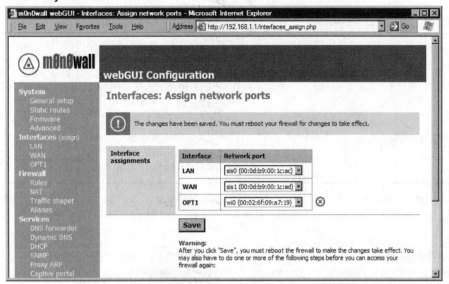

Before rebooting as prompted, you can configure the wireless card so it will be active after the reboot.

Click the newly created **Interfaces | OPT1** menu item and enter the options as described in Table 6.11. The **bold** values are different from the default settings.

Table 6.11 Wireless Interface Option Settings

Option Name	Value	Comments
Enable Optional 1 Interface	**Checked**	Check this to turn on the radio when m0n0wall is restarted.
Description	**WLAN**	Change this to WLAN (Wireless LAN) to have this appear elsewhere in the configuration pages instead of the less obvious OPT1 name.
Bridge With	none	Bridging should be off if you wish to use advanced features like captive portal.
IP Address	**192.168.2.1/24**	Enter a different, nonoverlapping, subnet other than your LAN interface.
Mode	hostap	Also known as "access point" mode
Service Set Identifier (SSID)	**630.Camerana**	How your network will show up when someone scans for access points. Assuming you want people to be able to find you, it's best to provide some contact info as the SSID, such as a street address, phone number, e-mail, or Web site address. For privacy reasons, you may not wish to divulge this information.

Continued

Table 6.11 Wireless Interface Option Settings

Option Name	Value	Comments
Channel	6	Scan from your wireless computer to see what channels are already in use and pick a different one. Be aware that adjacent channels interfere, so it's best to choose between channels 1, 6, or 11, or perhaps 1, 4, 7 and 11. Some countries allow higher channels.
Enable Wired Equivalent Privacy (WEP)	Unchecked	Enabling this too soon makes debugging bad settings much harder, so leave WEP disabled until you verify that everything is working okay.

The completed screen is shown in Figure 6.20. Click **Save** when you're done. Just one more page of settings before its time to reboot.

Figure 6.20 m0n0wall Wireless Configuration Settings Completed

After the wireless interface is configured and enabled, you need to set up DHCP to provide addresses for computers connecting via wireless. Click the **Services | DHCP** menu link and you'll now find two tabs to choose from: the *LAN* tab that was always there, and a new *WLAN* tab (or OPT1 if you chose not to rename it). Turn on the **Enable DHCP Server On WLAN Interface** setting and set the range as desired. For example, a range from 192.168.2.100 to 192.168.2.199 is analogous to the default provided via DHCP on the LAN interface. On this same page, you can also define any static IP-MAC mappings you'd like to enter (as described earlier in this chapter). Click **Save** one final time and you'll see Figure 6.21.

Figure 6.21 m0n0wall Wireless Interface DHCP Configuration

Choose the **Diagnostics | Reboot system** menu and click **Yes** to restart with your new settings. When it restarts, you should now have a working wireless network! Verify this by firing up your wireless computer and scanning for available networks. You should see the SSID you specified and your computer will have an IP address and appropriate parameters for DNS servers, gateway, and subnet mask.

However, you still can't use the Internet. For security reasons, the default firewall settings don't apply to new interfaces. You have to create a new rule that allows the WLAN interface to access the WAN interface (or LAN interface if you wish LAN and WLAN computers to share files, for example).

Click **Firewall | Rules** to see the LAN rule that already exists, as shown in Figure 6.22.

Figure 6.22 Default m0n0wall LAN Firewall Rules

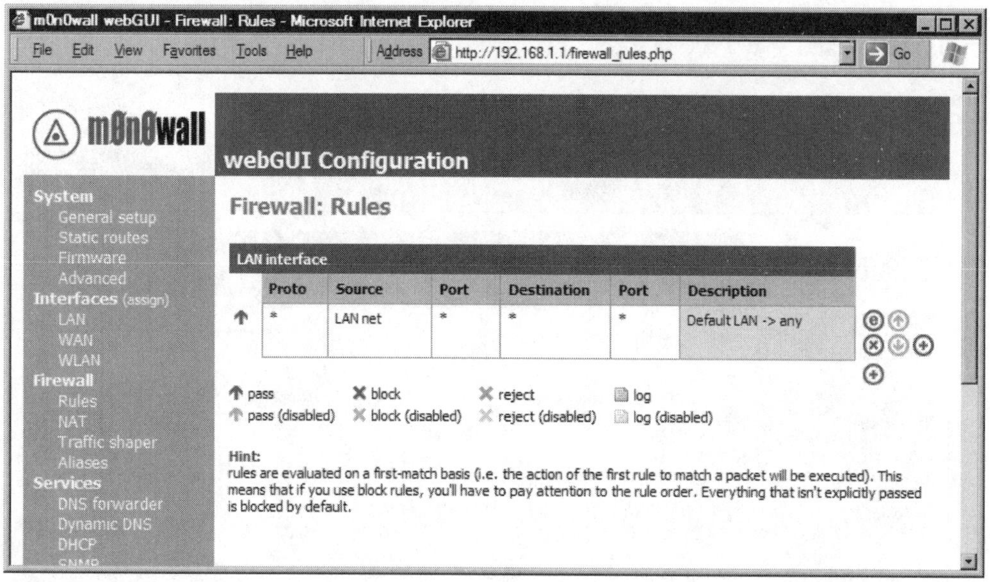

Another rule is needed to allow WLAN access. It should look just like the existing LAN rule, except replacing LAN with WLAN. Click the + (plus) sign on the bottom right below the existing table to open the new rule screen shown in Figure 6.23.

Figure 6.23 The New Rule Screen

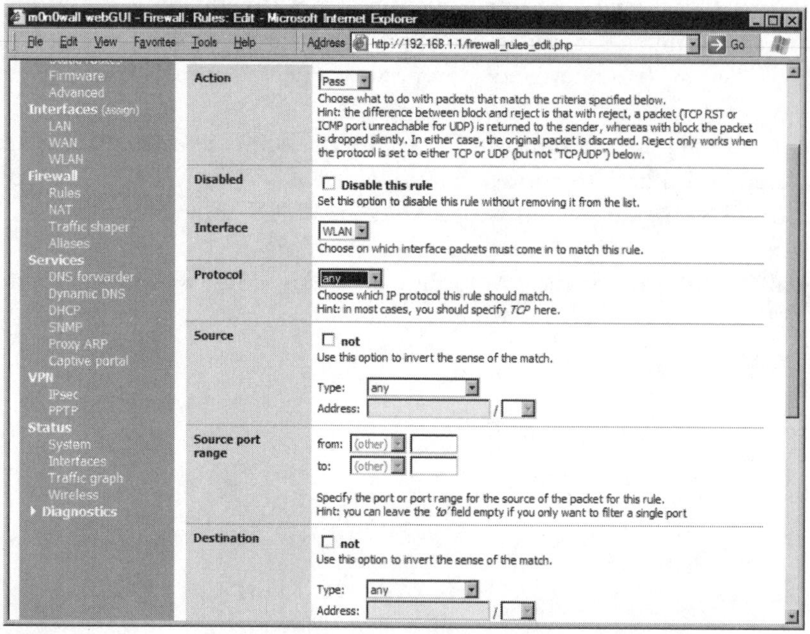

Because this is a broad rule, only three fields need to be changed.

1. **Interface** should be changed from WAN to **WLAN**.

2. **Protocol** should be changed from TCP to **any**.

3. Change **Source..Type** from any to **WLAN subnet**.

4. (Optional) Change **Description** to **WLAN -> any**.

Click **Save** to create the rule and then click **Apply Changes** to start the rule working. Figure 6.24 shows the results.

Figure 6.24 m0n0wall WLAN Access Rules

After making the changes, reboot the firewall. You should now be able to surf successfully. Congratulations, you have a working wireless access point that is extremely powerful and configurable!

Captive Portal—Requiring Sign On

Wireless access points are being installed in public spaces all over the world. Many of these are free for "appropriate use," where "appropriate" can vary dramatically. One convenient way to inform people about the appropriate use is to show them the rules before they start using the connection.

The common solution for this is to redirect any incoming requests to a Web page that you design and block all access until the user acknowledges the page by clicking a Continue button, or similar. This is commonly termed a "captive portal." Recent versions of m0n0wall provide captive portal support and in this section you'll learn how to turn it on. m0n0wall's captive portal is evolving quickly, so the following may not be exact, but the principles will remain unchanged.

Before you can configure captive portal, you need to:

- Decide which interface you wish to make captive. You can't be using bridged interfaces, so for example, any wireless radio you have must be separated onto its own subnet running DHCP (as described earlier in this chapter).

- Design an HTML page that will be displayed before the user can continue.

m0n0wall also supports authenticated sign-in using the popular RADIUS protocol, in case you only want to allow access to certain users. You'll need a separate RADIUS server to use that feature.

A minimal HTML page to use for the portal is:

```
<html><body>

<center>

<h1>Welcome to my Captive Portal</h1>

<form method="post" action="">

    <p>Click the Continue button to start surfing.</p>

    <input name="accept" type="submit" value="Continue">

</form>

</center>

</body></html>
```

Save this as a file—for example, captiveportal.html. Note that you cannot use images in this Web page as there is nowhere to store them in m0n0wall. When you're done with the HTML, click **Services | Captive Portal** to show the settings page. Change the settings as follows:

1. Check the **Enable Captive Portal** option.

2. Choose the desired interface—for example, **WLAN** (or **OPT1**).

3. Specify the HTML file to display for the portal page (for example, captiveportal.html)

4. Click **Save**.

The admin interface is exempt from the portal, so you'll be able to continue viewing the administrative pages. If you open another Web browser, however, and go to your favorite Web site, you should be greeted with the screen shown in Figure 6.25.

Figure 6.25 The Captive Portal Page in Action

By clicking **Continue**, the Web site you were trying to visit will be displayed. All other Internet access is blocked until the button is clicked, so if you're used to opening your e-mail program before opening a Web browser, you'll find that the e-mail fails to work until you acknowledge the sign-in page in a Web browser.

The captive portal configuration can be tweaked further. For example, you can

- Decide which Web sites, if any, should be allowed through the portal (for example, the Web site of a sponsoring company for a public hotspot). You can use this feature to add images to your captive portal page.

- Gather any computer MAC addresses that may bypass the portal (for instance, a Wi-Fi phone or other device lacking a browser interface that needs outside access). This functionality is sometimes referred to as a "whitelist."

These are entered under the *Allowed IP Addresses* and *Pass-through MAC* tabs of the **Services | Captive portal** page.

You can monitor use of your captive portal with the **Status | Captive portal** page. This will let you view what computers are currently using the system and the time of their most recent activity.

Finishing Off—Making a Backup

Last but not least, anytime you've configured m0n0wall to your satisfaction, make a backup of all the configuration settings. This is fast and simple. All the settings are saved into a single file in XML format. You can even edit the file with a text editor to make simple changes to most settings.

Click the **Download Configuration** button to save the settings, as shown in Figure 6.26. You may wish to rename the file from config.xml to m0n0config.xml or something even more meaningful to help retrieve it later. The same configuration file works across all m0n0wall platforms, which is handy when upgrading, say, from an old PC to a compact, quiet single board computer.

Figure 6.26 Back Up and Restore All m0n0wall Settings

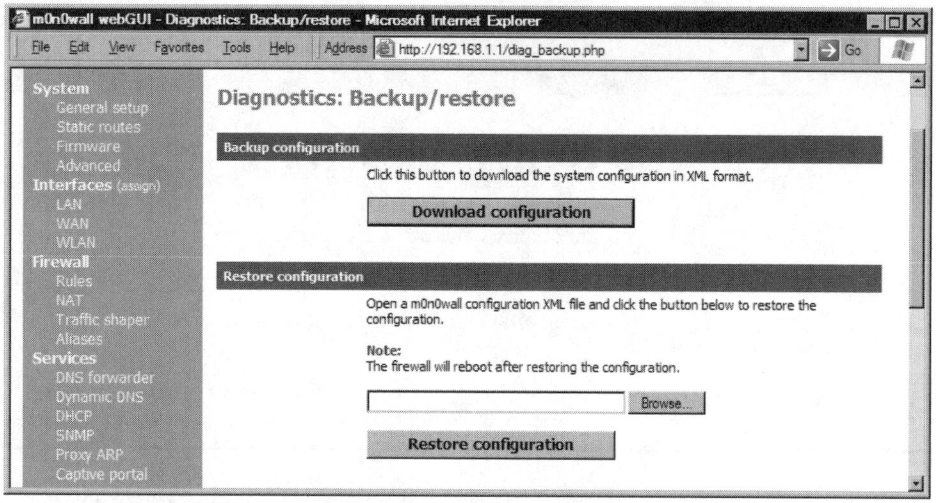

Under the Hood: How the Hack Works

m0n0wall is a fascinating firewall and access point distro. It is unique in design because it is based on FreeBSD, but all the configuration and administration is powered by a high-level PHP scripting language. PHP is designed for building Web pages, and is well suited to making the powerful configuration screens that m0n0wall provides. However, it is unusual to also use PHP to create and manage the myriad of files required by BSD's many programs that run behind the scenes to provide m0n0wall's rich functionality.

It's possible to get "behind the scenes" of m0n0wall with some special hooks provided for developers and the technically curious. For example, if you enter the URL: **http://192.168.1.1/status.php**, you'll see the output of many of the underlying BSD programs.

Another powerful tool is found at: http://192.168.1.1/exec.php. This allows you to execute any FreeBSD command and see the output, or to upload or download a file. Using these tools you can modify the underlying files for m0n0wall while it is running. More details can be found in the *hackers guide* on the m0n0wall Web site at http://m0n0.ch/wall/hackersguide.php.

Pebble—Powerful, Raw, Complete

The *Pebble* distro is the creation of Terry Schmidt, a talented member of NYCWireless. It is available at their Web site: http://nycwireless.net/Pebble.

The *Pebble* distro is a version of the Woody release of Debian Linux which has been stripped down and modified to run on a 64MB compact flash card. It stands apart from other distros in several key ways:

- The compact flash card is a (mostly) read-only disk while running. This allows more functionality and convenience for those familiar with Linux as many of the standard Linux commands are available and the configuration steps are almost identical.

- Up-to-date support for the latest wireless cards, particularly 802.11a and 802.11b/g cards which are often based on Atheros-designed chips.

- Includes nocat captive portal software (www.nocat.net) pre-installed, pre-configured, and ready for immediate use.

- It is based on Debian Linux and includes its excellent package distribution and update system: aptget. This allows for easy in-place security updates and feature additions.

- Broad-based wireless community support with many additional packages and enhancements available (including snmp and mesh-like support).

Here are some other interesting features of Pebble when used as an access point, or when connected to an access point.

- Network Address Translation (NAT)

- Dynamic Host Configuration Protocol (DHCP)

- Caching DNS forwarder (simplifies client configuration)

- SNMP agent for remote usage monitoring (optional add-on)

- Appropriate ISP support (DHCP client, PPPoE, PPTP)

- CD-ROM-only–based version is also available (boots and runs from a customized CD)

- Zebra and OSPF support

Pebble has many more advanced routing and firewall features that are beyond the scope of this chapter. See the Pebble Web site for more information.

Preparing for the Hack

Pebble is designed to run on single board computers using a compact flash card. However, it will also work on a standard PC using a compact flash to IDE adapter, or directly from a hard disk. This section describes installing an SBC version.

Every installation of Pebble is unique, with newly generated security keys and root passwords. Accordingly, the compact flash card (or custom CD-ROM) must be generated from a system running Linux. If you don't already have Linux installed on a computer, don't worry! This guide will tell you how to run a version of Linux directly from CD-ROM and then install Pebble—leaving your PC hard disk unaltered. You will need the following items:

- A standard PC with 256MB or more memory (or a Linux system), and a compact flash card writer

- A Knoppix CD—to turn your standard PC into a Linux machine without touching your hard disk

- A 64MB or bigger compact flash card

- A single board computer with a Prism or Atheros chipset wireless card

- (Optional) An SSH client terminal program. A good Windows program is SSH Secure Shell from www.ssh.com (look for the noncommercial "SSH Secure Shell for Workstations" download).

Some brands of compact flash cards work better than others, as Linux uses their built-in hard disk emulation in a different way from most photo cameras and similar devices. See the Pebble README file on the Pebble page at their Web site for specific recommendations.

Performing the Hack

The hardest part of installing Pebble is negotiating Linux to get it copied onto the compact flash card. If you're already familiar with Linux, then this is going to be easy and familiar! If Linux is not your specialty, don't worry, we'll cover all the bases and go step by step. Skip ahead to the "Downloading Pebble" section if you already have Linux running.

Creating a Boot CD and Starting Knoppix

Download Knoppix from the main Web site at http://knoppix.net or purchase a CD from many of the vendors who sell it for a small reproduction fee. It's a large 700Mb download, so it may be worth trying to use one of the advanced download tools like bittorrent that can do piecemeal downloads and automatically restart as needed.

Once you have the ISO image downloaded, burn a CD with it, as described earlier in this chapter in the section "Creating a CD-ROM from Windows."

Then boot your computer from the CD-ROM as described earlier in this chapter in the section "Booting from a CD-ROM and a Blank Diskette," except you don't need the blank diskette. When Knoppix has loaded, you'll see a screen similar to Figure 6.27.

Figure 6.27 The Knoppix Main Screen After Loading

Most of the Pebble installation will be done from the command line *shell*. This is done using Linux shell commands. First you need to open a Linux console or *shell,* where you can enter the commands. In the Knoppix graphical environment, you can do this by performing the following steps:

1. On the bottom of the screen toolbar, click the icon that looks like a small black box or computer screen.

2. After a pause, a large black box will appear with a shell prompt: knoppix@ttyp0[knoppix]$.

3. In the box, type **su** and then press **Enter**. This will change the prompt to root@ttyp0[knoppix]$.

Figure 6.28 shows this process, but the black background in this and subsequent images has been reproduced as white for clarity.

Figure 6.28 Starting the Knoppix Command Line Shell

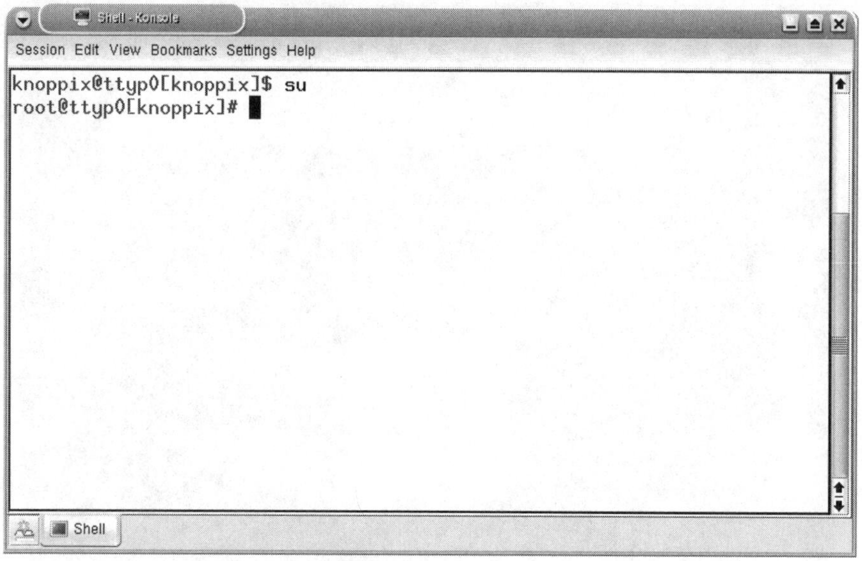

Now you have a Linux console and you're running as *root* (also known as *super user*). This gives you the permissions necessary to install Pebble.

Configuring the Compact Flash Reader/Writer

One of the trickiest parts of this installation is to verify that the compact flash reader (and writer) is working correctly and to determine its device name.

You can determine if your compact flash reader has been recognized by first using the command: **ls /proc/scsi/usb-storage-0**.

If nothing happens when you do this, then your compact flash reader is not recognized. If it is recognized, then you should see a number. Use that number for the next command. For example, if the number is 2, then issue the command: **cat /proc/scsi/usb-storage-0/2**.

This should display some information about the USB card reader you are using. These commands and the results are shown in Figure 6.29.

Figure 6.29 Linux USB Storage Information

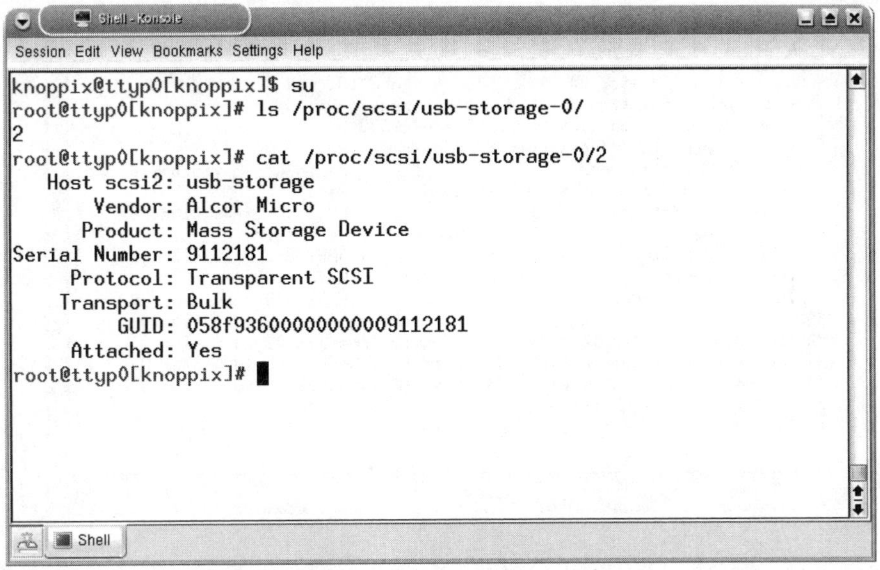
Shell - Konsole
Session Edit View Bookmarks Settings Help
knoppix@ttyp0[knoppix]$ su
root@ttyp0[knoppix]# ls /proc/scsi/usb-storage-0/
2
root@ttyp0[knoppix]# cat /proc/scsi/usb-storage-0/2
Host scsi2: usb-storage
Vendor: Alcor Micro
Product: Mass Storage Device
Serial Number: 9112181
Protocol: Transparent SCSI
Transport: Bulk
GUID: 058f93600000000009112181
Attached: Yes
root@ttyp0[knoppix]#
Shell

If you are using a multiformat card reader, your reader may not be recognized. A possible fix for this is to modify the Knoppix boot procedure. Instead of letting it boot from CD-ROM with the default settings, enter the following at the prompt: **knoppix max_scsi_luns=6**.

Then repeat the steps presented previously and see if it is now recognized. If it isn't, you'll need to do some further online research. The Knoppix site has well-supported message boards with a good search facility. Searching for, say, "6-in-1," or the card reader's manufacturer name may provide the necessary hints.

If the reader is recognized, you'll still need to determine what Linux calls it, its so-called *device name*. Type the command **dmesg | more** (the middle vertical line is usually found above the **Enter** key). Use the Spacebar to skip through the information one page at a time until you see something about "SCSI Emulation USB Mass Storage Devices."

Figure 6.30 shows the output for a "no brand" 6-in-1 reader.

Figure 6.30 USB Storage Device Name Information

```
Shell - Konsole

Session Edit View Bookmarks Settings Help

hub.c: new USB device 00:11.2-1.1, assigned address 3
scsi2 : SCSI emulation for USB Mass Storage devices
usb-uhci.c: interrupt, status 3, frame# 1339
    Vendor: Generic    Model: USB SD Reader      Rev: 1.00
    Type:    Direct-Access                       ANSI SCSI revision: 02
    Vendor: Generic    Model: USB CF Reader      Rev: 1.01
    Type:    Direct-Access                       ANSI SCSI revision: 02
    Vendor: Generic    Model: USB SM Reader      Rev: 1.02
    Type:    Direct-Access                       ANSI SCSI revision: 02
    Vendor: Generic    Model: USB MS Reader      Rev: 1.03
    Type:    Direct-Access                       ANSI SCSI revision: 02
Attached scsi removable disk sda at scsi2, channel 0, id 0, lun 0
Attached scsi removable disk sdb at scsi2, channel 0, id 0, lun 1
Attached scsi removable disk sdc at scsi2, channel 0, id 0, lun 2
Attached scsi removable disk sdd at scsi2, channel 0, id 0, lun 3
sda: Unit Not Ready, sense:
Info fld=0x0, Current 00:00: sns = f0  2
:█

    Shell
```

Here you can see that the USB CF Reader is the second USB storage device. When you scroll further down, you'll find more information about the corresponding devices. These include the device name. For example, almost right at the bottom of Figure 6.30 on the left is the device name sda. This suggests that the USB SD Reader has the device name of /dev/sda. As the next device is the CF reader, it's likely that it will have the name /dev/sdb. The way to determine if this is the correct name is to insert a CF card and try and format it. This is described next.

Formatting the Compact Flash Card

Before you copy Pebble to the CF card, you need to partition and format it. Most likely the card is already formatted in so-called *FAT* format. That partition needs to be deleted and replaced with a Linux *ext2* partition. First, insert the CF card into the reader. Then run the command (substituting your device name, of course):

```
cfdisk /dev/sdb
```

where you should replace /dev/sdb with whatever device name you've determined is appropriate for your compact flash drive. For example, /dev/sda is a common choice if you have a single format compact flash reader and no other USB storage devices on your system.

WARNING: HARDWARE HARM

It's vital to verify which device corresponds to your CF card, because you can easily overwrite your (primary!) hard disk or other storage devices like USB memory keys with this command. This can be especially difficult in Linux (Knoppix), so be careful! We suggest you remove any nonessential storage devices before attempting to write to your CF card. A good double-check is to ensure that the size of the device as reported by the various utilities, such as cfdisk, matches the known size of the compact flash you're using.

The cfdisk utility is a text menu program that shows you the layout of your "disk drive," (that is, the compact flash card) and lets you modify it. It's analogous to the fdisk program in DOS, but easier to use. You'll need to perform the following steps:

1. **[Delete]** any existing partitions—there should only be one or zero. If there are more, you most likely have the wrong device.

2. Use **[New]** to create a new partition.

3. Select **[Primary]** as the partition type.

4. Make the size take up the entire card by pressing **Enter**—be sure the size matches the size of the compact flash card.

5. Make the card [Bootable].

6. Select **[Quit]** and save your changes when prompted.

Figure 6.31 shows the cfdisk display when you've finished preparing your compact flash for its first use.

Figure 6.31 Newly Partitioned Compact Flash

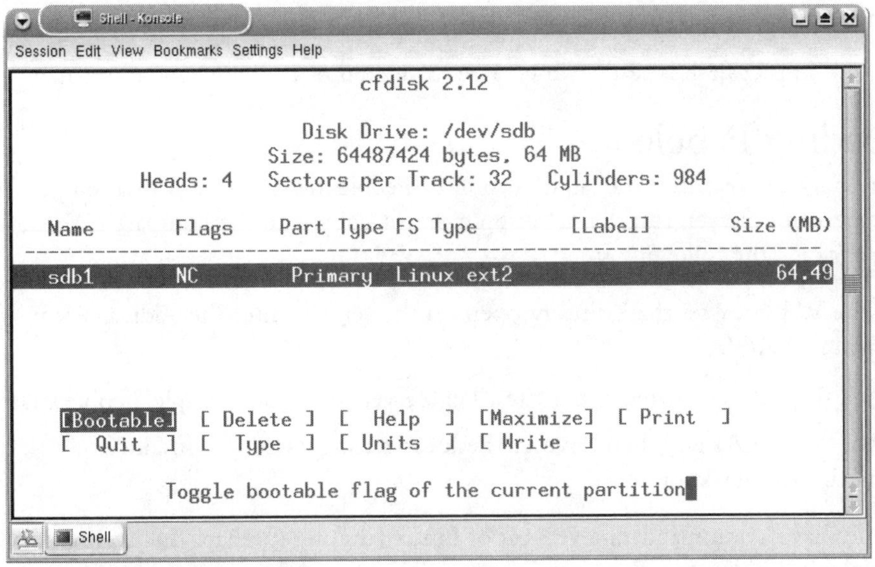

Once the card has been partitioned, you need to format it. Use the command (substituting your device name): **mkfs.ext2 /dev/sdb1**.

Note that there is now the number "1" (one) on the end of the /dev/sdb device name. This stands for *partition number one*. The results of the mkfs.ext2 command for a 64MB CF card are:

```
mke2fs 1.35-WIP (31-Jan-2004)
Filesystem label=
```

```
OS type: Linux
Block size=1024 (log=0)
Fragment size=1024 (log=0)
15744 inodes, 62972 blocks
3148 blocks (5.00%) reserved for the super user
First data block=1
8 block groups
8192 blocks per group, 8192 fragments per group
1968 inodes per group
Superblock backups stored on blocks:
        8193, 24577, 40961, 57345


Writing inode tables: done
Writing superblocks and filesystem accounting information: done


This filesystem will be automatically checked every 30 mounts or
180 days, whichever comes first.  Use tune2fs -c or -i to override.
```

Now your compact flash is ready. Time to get a version of Pebble to install on it.

Downloading Pebble

We'll use Knoppix to download the latest version of Pebble. It will place it in a temporary "disk" in memory where you can then run the install program to copy it to the compact flash card. To download Pebble, perform the following steps:

- In the Web browser that's already open on the screen, enter the address **www.nycwireless.net/Pebble**.

- Click the link to download the latest Pebble version—for example, Pebble.v41.tar.bz2.

- Choose **Save As** and then **Save** to the location suggested (for instance, /ramdisk/home/knoppix).

If you only have a command-line version of Linux running (perhaps due to memory constraints), you can download Pebble using the **wget** command. For example:
wget http://www.nycwireless.net/Pebble/Pebble.v41.tar.bz2
When the download completes, you'll have the latest Pebble and you're ready to install it.

Copying Pebble to the Compact Flash

Run the following commands to prepare for the Pebble installation script:

1. **mkdir /mnt/cf** Creates a *mount point* for the compact flash card.

2. **mkdir Pebble** Creates a directory to untar Pebble into.

3. **cd Pebble** Changes to the new Pebble directory.

4. **tar --numeric-owner –jxvf ../Pebble.v41.tar.bz2** Untars the Pebble files.

If there isn't enough RAM on your PC, you will see errors in step 4. If it all works successfully, you'll see a list of filenames scrolling by, ending with the command prompt. If there are errors, there will be messages about running out of space for all the remaining files. (If you start seeing these, you can press **Ctrl + C** to stop the process.)

Now you can run the **Pebble.update** script. It will copy the Pebble files to the compact flash and then write the appropriate passwords and security keys. Issue the command **./Pebble.update**, which will run and ask the questions shown. These are answered in bold.

```
Welcome to the Pebble Linux installation script


Where is the Pebble installer (this) directory? (default=/mnt/Pebble):
```
<enter>
```
Which device accesses the compact flash? (default=/dev/hde):
```
/dev/sdb
```
Which directory should I mount the flash card to? (default=/mnt/cf):
```
<enter>
```
Which module? Enter 1 for pcmcia, 2 for net4501, or 3 for net4521/net4511 (default=2):
```
3

```
Saving defaults to .Pebble.config



Configuration completed:
----------------------


Installer directory: /mnt/Pebble

FlashCard Device:     /dev/sdb
Will be mounted on:   /mnt/cf


Installation mode:    3


If this is good hit RETURN, otherwise hit CTRL+C
```
<enter>

After you press **Enter**, the script will copy the necessary files. It will then prompt you for a *root password*. This password is needed to modify the Pebble configuration after it's running on your SBC, so choose it carefully and write it down since you'll need it soon.

Booting Pebble

Connect a serial port as described previously in the section "Starting Your SBC." For the Soekris board described here, you'll need to start your terminal program with a baud rate of 19200. Then you can use **Ctrl + P** (from Table 6.6) to interrupt the boot sequence and change the baud rate to 9600 using the command **set conspeed 9600**.

At the same time, you may as well add the command **set bootdelay 2** to shave a few seconds off the default boot delay time of 5. Disconnect the power, reset the baud rate on your serial terminal program to 9600 and then reconnect the power. Now Pebble should start booting and finally offer a login prompt. Log in using the username *root* and the password you defined during setup, as follows:

```
Debian GNU/Linux 3.0 Pebble ttyS0


Pebble login: root
Password:
Pebble Distro - smallish read-only.  < http://www.nycwireless.net/Pebble/ >
/Pebble.README - readme file
/usr/local/sbin/remountrw - remount the file system read-write
/usr/local/sbin/remountro - remount the file system read-only
/usr/local/sbin/fastreboot  - do a fast reboot, and remountro
/usr/local/sbin/remove.docs - removes docs files, and other excess baggage
Pebble:~#
```

The only thing left to do is to turn on your radio card and you'll have a full-featured access point with captive portal functionality.

Configuring Pebble

Pebble is configured by default to use DHCP on the first Ethernet port, eth0, to find its connection to the outside world and then act as an Access Point with captive portal for wireless clients. It has the settings shown in Table 6.12 for the wireless clients.

Table 6.12 Default Pebble Wireless Internet Settings

Setting	Value
Gateway	192.168.89.1
Subnet mask	255.255.255.0 (/24)
DHCP address range	192.168.89.10 to 192.168.89.250
DNS server	192.168.89.1

Any other Ethernet ports are disabled.

Using a Prism-Based Radio

Pebble supports many common radio cards via the hostap drivers. The full list of cards supported can be found at the hostap driver pages and the Linux Wireless LAN Howto at http://hostap.epitest.fi/ or www.hpl.hp.com/personal/Jean_Tourrilhes/Linux/Wireless.html.

Your card must be listed as supporting "Host AP mode" for you to use the card as an access point (versus using it as a client to connect to some other access point).

By default, the radio support is commented out. Here are the steps to enable a HostAP supported card:

- Log in to Pebble via the serial console.

- Type **remountrw** to enable read/write mode and allow changes to be saved to the CF card.

- Type **editor /etc/modules** to start a simple full screen editor.

- In the editor, find the line **#hostap_pci** and remove the **#** (comment) symbol from the start of the line.

- Save the file with **Ctrl + O** (the letter "o," not the number zero), then press **Enter** and **Ctrl + Q** to quit.

- Type the command **fastreboot** to save your changes and restart the computer.

Using an Atheros-Based Radio

Atheros chipset radios are popular for 802.11a and 802.11g solutions. Pebble provides support for these, but it is not enabled by default. Here are the steps needed to use an Atheros radio:

- Log in to Pebble via the serial console.

- Type **remountrw** to enable read/write mode and allow changes to be saved to the CF card.

- Type **editor /etc/modules** to start a simple full-screen editor.

- In the editor, add a new line: **ath_pci**.

- Save the file with **Ctrl + O** (the letter "o," not the number zero), then press **Enter** and **Ctrl + Q** to quit.

- Similarly, modify the file **/etc/network/interfaces**, replacing **wlan0** with **ath0**.

- Similarly, modify the file **/etc/default/dhcp**, replacing **wlan0** with **ath0**.

- Type the command **fastreboot** to save your changes and restart the computer.

Note that only lines without a leading "#" symbol need to be modified. The # is used to indicate comment lines (which are ignored by the executing program), and are often used to disable a certain setting.

Changing the SSID

By default, Pebble assigns the SSID freenetworks.org. If you wish to change the SSID or channel, follow these steps:

1. Log in to Pebble via the serial console.

2. Type **remountrw** to enable read/write mode and allow changes to be saved to the CF card.

3. Type **editor /etc/network/interfaces** to start a simple full-screen editor.

4. Use the arrow keys to move to the text FreeNetworks.org and replace it with your preferred SSID.

5. Save the file with **Ctrl + O** (the letter "o," not the number zero), then press **Enter** and **Ctrl + Q** to quit.

6. Type the command **fastreboot** to save your changes and restart the computer.

After the reboot, your new SSID should now be visible to wireless clients.

Adjusting Other Pebble Settings

You now have a good working system, but you may want to make additional changes. Configuring a Linux network can be a complex and rewarding task. Fortunately, there are many resources on the Internet to help. The How-To articles are particularly useful. They're written in a uniform style and tend to be comprehensive, yet are tailored for Linux newcomers. Each flavor of Linux is a little different, so search for articles that explain the Debian way of configuring files. Pebble is based on the version of Debian called "Woody."

Under the Hood: How the Hack Works

Pebble is a stripped down Linux distro that takes advantage of the relatively small size of Linux and the ever-growing size and dropping prices of compact flash memory. By default, it isn't possible to simply use a compact flash card as a hard drive replacement. This is where some of the magic of the Pebble distro becomes apparent.

Like most operating systems, Linux uses its disks heavily for writing log files and updating miscellaneous status files. Over a period of time this can "wear out" the permanent memory spaces in the compact flash. Also, if Linux is writing to its file system and there is a power interruption, the disk can be left in an unstable state.

Most distros deal with this problem by using the compact flash only at startup time to copy the entire contents into a *memory disk*. This is inherently inefficient, of course, as all the files are stored twice. Pebble resolves this by mounting the compact flash in *read-only* mode and using the files directly from the compact flash as needed, and writing temporary files and other system state to the memory disk. This sounds simple, and in theory it is. However, in practice, Linux doesn't normally split itself between read-only and read-write media, so getting the details right and having it all work reliably is an admirable feat.

The readme file available on the NYCWireless Web site is a treasure trove of information. Be sure to read it closely as you start to explore the many powerful features of Pebble.

Chapter 7

Monitoring Your Network

Topics in this Chapter:

- **Enabling SNMP**

- **Getif and SNMP Exploration for Microsoft Windows**

- **STG and SNMP Graphs for Microsoft Windows**

- **Cacti and Comprehensive Network Graphs**

Introduction

If you build a wireless network for personal use, you'll quickly know if there are critical problems with it since you're the only one using it. Likewise, if its performance lags over time as you use it more (e.g., streaming video via wireless to the TV in your den slows down), you'll notice that too and can plan upgrades as needed.

However, if you're using some of the advanced equipment and techniques suggested in this book, chances are your network will be used by many others. If you don't live in the neighborhood where the network is deployed, perhaps you won't be using it at all. So when problems happen, and they will happen, you won't know until someone calls with the question: *Is the network down?* And when they do call, you won't have any historical information to guide your diagnosis. This is especially vital if your network consists of multiple Access Points linked via various means.

For example, one SoCalFreeNet network in San Diego has multiple Access Points linked together via various 802.11a backhaul radios. If someone contacts us with a problem, we can check the graph for each Access Point and the backhaul links to see if they've been passing traffic. We also have *stacked* graphs that show the cumulative bandwidth from node to node versus the total traffic going through the main Internet DSL feed. These graphs help us pinpoint a specific link problem, or identify large traffic mismatches caused by, say, virus or worm traffic trying to get out through the firewall but getting dropped instead.

Having simple traffic graphs can also help with traffic capacity management, both by you and your users. For example, if your users can easily discover that the system is very busy each night at 8P.M., but relatively quiet at 8A.M., they'll probably decide to do their large, bandwidth-intensive downloads after getting up in the morning instead of waiting for them at night.

In this chapter, we'll talk about some different monitoring systems that provide graphic views of your equipment and its operations. Some run directly on a desktop PC to provide immediate data, while others run on a server to provide historical charts as well as up-to-date information. These tools fall short of full-blown monitoring systems because they don't specifically target management concerns like configuration, security, fault detection, or account management. Nor are they proactive monitoring systems that attempt to automatically detect failures and send e-mail or pager notifications, or try to correct the problems. However, they are a rich a source of useful information that will help greatly with the day-to-day operations and tuning of the network.

All the monitoring tools we discuss in this chapter use an industry standard protocol called Simple Network Management Protocol (SNMP). This protocol has two pieces: network devices that provide status using SNMP, and SNMP applications that gather and present the data. So, for example, when monitoring a wireless network, you will have at least one Access Point with SNMP support and then, say, a PC running an SNMP monitoring program that regularly polls the devices for their status. Or the monitor could be a Web server with a database of results that generates Web pages as needed to view the various statistics.

Enabling SNMP

Most wireless devices support a monitoring system called SNMP. This protocol provides a standard mechanism for querying a device for many standard parameters such as the system name and manufacturer. However, fortunately for our needs, they also report the network interfaces and various statistics about the interfaces such as the number of bytes transmitted and received. Plus, in more advanced usage, you can also use SNMP to configure devices, though few consumer devices support that and we won't be delving that far into SNMP here.

Preparing for the Hack

In preparing for the hack, you'll first need to determine if your network devices support SNMP monitoring (most current consumer wireless equipment supports basic SNMP monitoring). SNMP has evolved since it was created and exists in versions 1 through version 3. All you need for basic monitoring is version 1. Linux-based systems, such as Pebble described in Chapter 6, may require the installation of appropriate SNMP tools, such as NetSNMP. Newer versions provide greater support for secure access, which is important if you're using SNMP to modify settings on your device, but less important for gathering basic statistics via a read-only connection, as described in this chapter.

Performing the Hack

To use the tools described in the rest of this chapter, you must first enable SNMP on the device you wish to monitor. Figure 7.1 shows the SNMP setup screen for the m0n0wall firewall software described in Chapter 6. Figure 7.2 shows the SNMP configuration for a typical consumer Access Point.

Figure 7.1 Enabling SNMP in m0n0wall

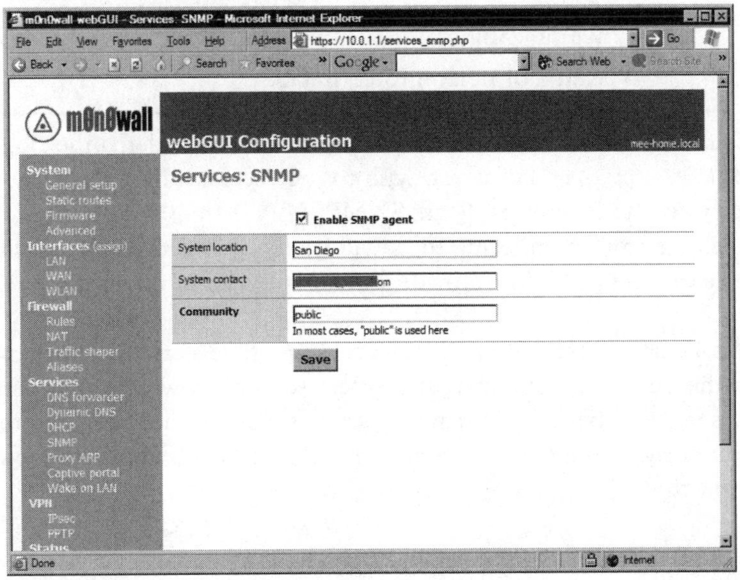

Figure 7.2 Enabling SNMP in D-Link AP

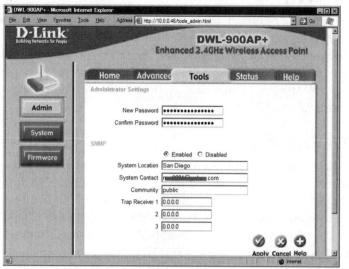

The three items usually needed for SNMP configuration on the device are described in Table 7.1.

Table 7.1 Common SNMP Device Settings

Setting Name	Explanation
Community	The "login" name to be used by SNMP tools to query this device. The commonest name is public.
System Location	A short description of where this device is located—e.g., first floor wiring cabinet.
System Contact	Name of person to contact.

The most critical setting is the Community name, which is considered the "login name" for the device. This is usually set to public, but if you wish to hide access more effectively, you could choose a different name. However, in its simplest form, SNMP V1.0, there is no security for this login name, so anyone with simple network monitoring tools will be able to see the Community name whenever you monitor it. Later versions of SNMP provide an encrypted login that is more secure from eaves-dropping.

The two System Location and System Contact settings are less critical for a small network. Chances are you're the only one monitoring the system so you know whom to contact. Similarly, the number of devices is likely to be so small that you know the location. These are provided for larger networks where there may be hundreds of devices that are automatically monitored by sophisticated network management tools.

When you enable SNMP monitoring for your network device, you are also enabling SNMP access for anyone on your network. Although this information is typically read-only and they cannot cause mischief by modifying your settings, some devices provide a lot of statistical and network specific information via SNMP that could be used to quickly gain detailed information about your network inappropriately. How much you worry about this will depend on how you're using your network.

Once you've enabled SNMP, you're all set to go with the tools described in this chapter. The first, Getif, is a good tool for confirming basic device functionality and configuration.

Under the Hood: How the Hack Works

When you enable SNMP on your device, you are telling it to listen on port 161 for requests from an SNMP query tool. These requests consist of the login information and an OID (object identifier), which specifies exactly what piece of information is needed. These OIDs are in turn listed together in groups called MIBs, or Management Information Bases. There are standard MIBs that contain OIDs for common requests such as interface numbers or packets sent or received, and there are various extension MIBs for specific areas like wireless. These allow you to query specific items like the current SSID setting, or the number of computers currently associated with an AP. Often, a manufacturer-specific MIB, such as Cisco's wireless extensions, is adopted by other vendors and it becomes a pseudo-standard.

Fortunately, the values that provide the most useful monitoring information are well standardized, so most devices will respond to the standard OIDs we'll be using later in this chapter.

Table 7.2 lists some resources on the Web to help you further explore the vast world of SNMP-based network monitoring tools.

Table 7.2 SNMP Resources

URL	Description
www.snmplink.org	Has links and information about SNMP and MIBs; also has a good Tools section with links to useful programs.
www.snmp4tpc.com	Acronym stands for SNMP For The Public Community. More PC-focused than most SNMP information. A good source of tools and information.
www.mibdepot.com	Has a very large collection of MIBs; a good place to find support for your specific device.

Getif and SNMP Exploration for Microsoft Windows

Microsoft Windows has long had its own built-in performance monitoring tools which are not based on SNMP. Perhaps this is why there are few good free tools for monitoring SNMP devices that run on Windows. However, as this is often the most convenient platform to start with, we will begin with a simple but powerful SNMP monitoring tool called Getif.

Getif is most useful for exploring a new device. With it, you can see what standard OIDs (queries) it supports. As you become more comfortable with the world of SNMP, you can load device specific MIBs into Getif and explore the device with the full text description of each OID. This is handy when trying to find that elusive OID that provides just the right information you need.

It will also do the simple graphing of a single device. However, it is limited to one graph at a time, so while it's good for a quick exploration, it is not as useful for monitoring multiple devices (or OIDs) at once.

Preparing for the Hack

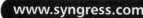

To use Getif, you'll need a computer running Microsoft Windows and the Getif Zip file. The author of Getif, Philippe Simonet, does not provide a Web site to download the file, so you'll need to simply do a search for "getif snmp" to find it. The download location with the most support and documentation is www.wtcs.org/snmp4tpc/getif.htm.

After you download the file, unzip it and then double-click the setup.exe program. Answer the usual questions about where you'd like it installed and you're ready to start!

Performing the Hack

Getif runs as a single multitabbed window. Figure 7.3 shows Getif's opening screen. It's a little daunting at first, but don't worry, we only need a small subset of the features to start graphing the network.

Figure 7.3 The Getif Opening Screen

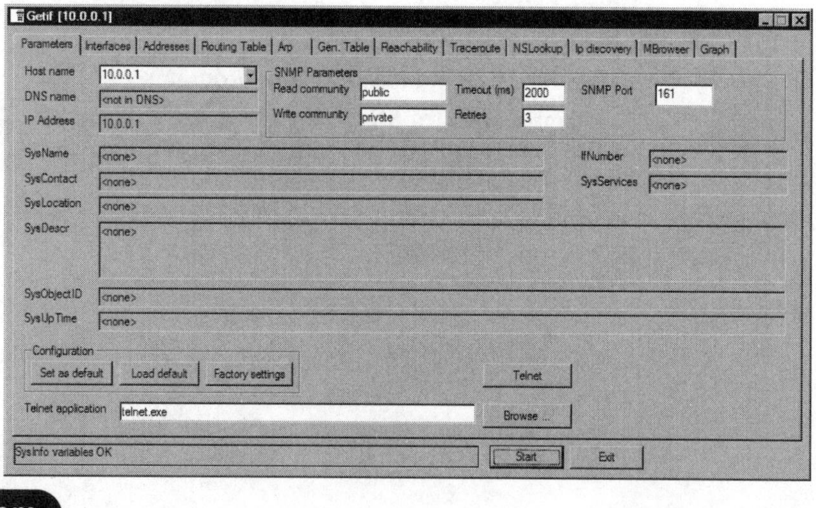

The first entry to fill in is the Host Name field. It is shown in Figure 7.3 with an IP address of 10.0.0.1 (the m0n0wall firewall is used as an example in this section). The Read Community field is set to "public". This corresponds to the value shown in Figure 7.1 and is the default value for a device, unless you changed it. Once these two settings are correct, you can click the Start button. If Getif successfully communicates with the device, the line of text at the bottom will read "Sysinfo variables OK", as shown.

Other devices may show more information—for example, the D-Link 900AP+ configured in Figure 7.2 will display information as shown in Figure 7.4 when you enter its IP address and click Start. Notice the SysName, ifNumber, and SysServices fields have been filled in along with some other data.

Figure 7.4 Getif Query Results from D-Link 900AP+

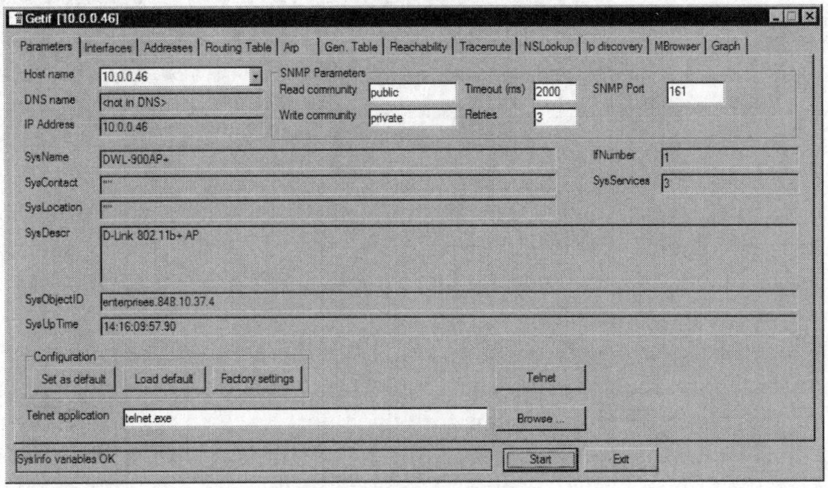

Once you have basic SNMP connectivity with the device, you're ready to begin monitoring.

Retrieving Device Interface Information

The next Getif tab is labeled **Interfaces**. Click this and you'll see two empty white boxes. Now click the **Start** button and it will query your device for what network interfaces it supports and replace the empty boxes with (potentially) several rows of data. Figure 7.5 shows the interfaces reported by m0n0wall.

Figure 7.5 m0n0wall Interfaces Reported by SNMP

A total of seven interfaces are shown. The last three, ppp0, s10, and faith0 are all shown as down in the admin and oper columns. If your m0n0wall system is running slip or ppp, you may see different results here. Interface number 4 is the standard local loopback interface at 127.0.0.1 and can usually be ignored.

The first three interfaces are the most interesting. The Ethernet interface names are sis0 and sis1. Other systems might report eth0 and eth1. These interfaces correspond to the local and WAN Ethernet ports on the m0n0wall device. A clue for which port is which is provided by the IP address column. This column shows that one interface is 10.0.1.1 and the other interface is 69.17.112.245 (the static IP of the WAN Internet connection). Therefore, in this example, sis0 is likely the local Ethernet port and sis1 is likely the WAN Ethernet port. The very first interface is wi0. This corresponds to the wireless radio card in the m0n0wall running at IP 10.0.0.1. On Linux-based systems, this would likely appear as wlan0 or ath0.

What have we achieved so far? Quite a lot! We're remotely querying our router, m0n0wall in this case, and seeing all the interfaces available along with some basic data about them. Be sure to use the horizontal scroll bar to see what other information is available. Some devices will report the Medium Access Control (MAC) address (sometimes referred to as the "Hardware" or "Ethernet" address) in the phys column, along with the corresponding hardware vendor.

Exploring the SNMP OIDs

So far so good, but what we really want to see is some interface statistics—for example, how much traffic is flowing through each port? To find that information, we need to explore the *MIB tree* for the device.

1. Click the **MBrowser** tab, then expand the following entries by clicking the **plus (+) sign** next to them:

```
iso

org

dod

internet

mgmt

mib-2

interfaces
```

2.   Click the word interfaces (instead what should now be a minus sign "–") sign next to it so that it's highlighted.

3.   Click the **Start** button. The white area immediately below should fill with entries. This is shown in Figure 7.6.

Figure 7.6 Browsing the m0n0wall MIB Tree to Find Interface Statistics

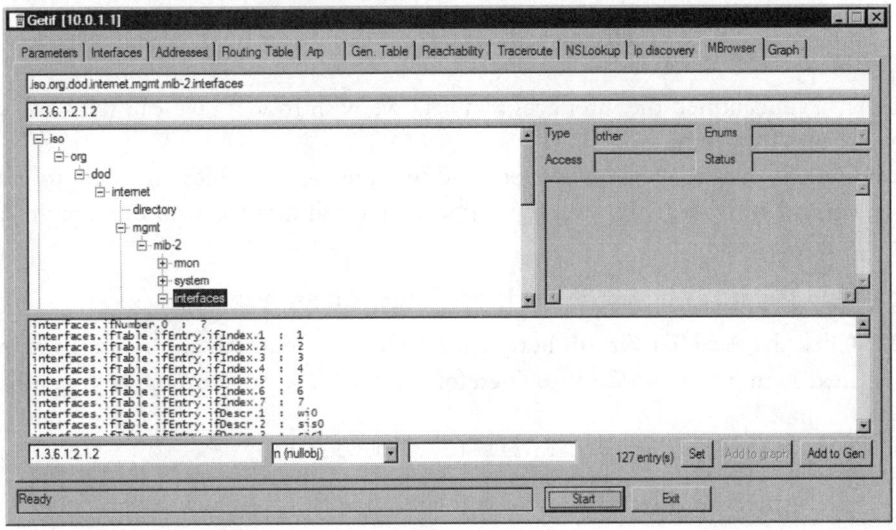

4.   We're almost done. In the bottom window, scroll down until you find the line that begins:

```
.interfaces.ifTable.ifEntry.ifInOctets.1
```

This shows the hierarchy of the MIB tree starting at interfaces (.interfaces), stepping through a table of all the interfaces (ifTable), then displaying each individual interface entry (ifEntry), followed finally by a specific value for that interface, reported as the number of incoming octets of data (ifInOctets). To the right of that is the actual number of octets received so far.

If you click other items in this lower window, the upper window will update and more information will appear in the grey box to the side. Figure 7.7 shows these details.

Figure 7.7 Amount of Data Received on Interface 1

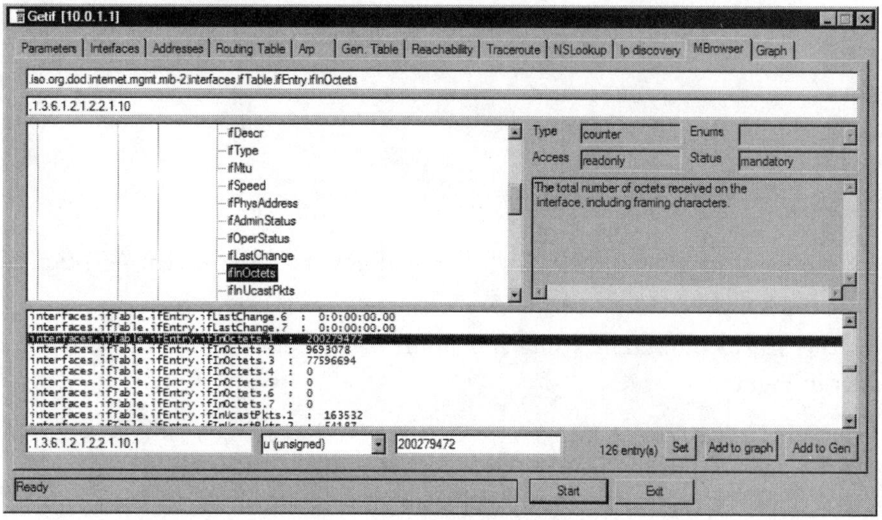

Graphing the Data

Now that we've identified the interfaces and data we wish to view, we can tell Getif to build a graph to show what is happening over time.

Continuing from the previous section, find the interface variables you wish to graph. For example, you might wish to show all the traffic data for all interfaces on one graph. To do this, perform the following:

1. Find the data you want in the lower white window pane.

2. Click the **Add To Graph** button for each line. Getif will automatically move down to the next item when you do this. Therefore, if you click Add To Graph three times, and then find the line

        .interfaces.ifTable.ifEntry.ifOutOctets.1

    and again click three times, you will end up with six elements being graphed.

3. Select the **Graph** tab at the top.

4. Click **Start** and the graph will begin plotting. Figure 7.8 shows a similar graph that has been running for a while. In the middle of the run is a large and then small bump corresponding to first a download and then an upload speed test.

Figure 7.8 Getif Graph of m0n0wall Firewall Traffic

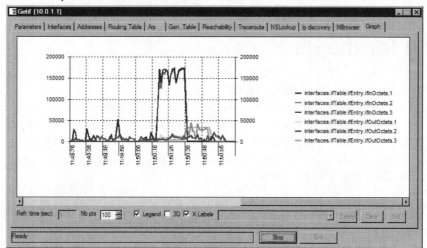

Under the Hood: How the Hack Works

The Getif program is doing quite a few things behind the scenes to make this as simple as possible, as will become clear in later sections of this chapter.

First, the opening Parameters tab and the adjacent Interfaces tab have some "canned" SNMP queries which use known OIDs from a standard MIB to fill the screen. This is a convenient shortcut to browsing the MIB tree to find individual values. One of the reasons the m0n0wall and D-Link devices returned different results for SysName and other values is that there is no strict standard for these values, so the "canned" queries worked better for the Linksys device than the (FreeBSD-based) m0n0wall firewall.

The MBrowser tab uses a precompiled MIB which contains all the OID numbers as well as corresponding descriptions of each value. It displays this in hierarchical tree form to make it easier to browse the data. When you click Start in the MBrowser tab, it "walks" the OID tree and queries for the OID values below that point. This also includes filling arrays of values, like data for each interface.

Finally, the graphing function automatically queries the device with the OIDs specified at the requested interval and then charts the results.

STG and SNMP Graphs for Microsoft Windows

STG, or SNMP Traffic Grapher, is a tool built with a single simple purpose: plot two SNMP OID values onto a graph. It's simple, effective, and does its job well with minimal hassle. You can also run it more than once so you have multiple graphs displayed simultaneously.

Unlike Getif, STG does not know anything about MIBs. Either the two default values it has predefined will work, or you'll need to use something like Getif to determine which OIDs you need to provide. This section builds on the previous Getif section and will step you through using STG to generate useful monitoring graphs.

Like Getif, STG does not show historical data, though it will log data in a text file for later analysis with some other program.

Preparing for the Hack

STG runs on any version of Windows 98 and later, including Windows XP. It can be downloaded from the author's site at http://leonidvm.chat.ru/.

Unzip the downloaded file and follow the instructions in the readme file if you're using an older version of Windows since some extra DLLs may be required. If not, you can run **stg.exe** directly from a command prompt (or select **Start | Run**), with no installation or setup process required.

Of course, you'll also need an available SNMP device to query. This was discussed previously in this chapter in the "Enabling SNMP" section.

Performing the Hack

Perform the following steps:

1. Start **stg.exe**. You will see an empty graph, as shown in Figure 7.9.

Figure 7.9 STG Waiting for Settings

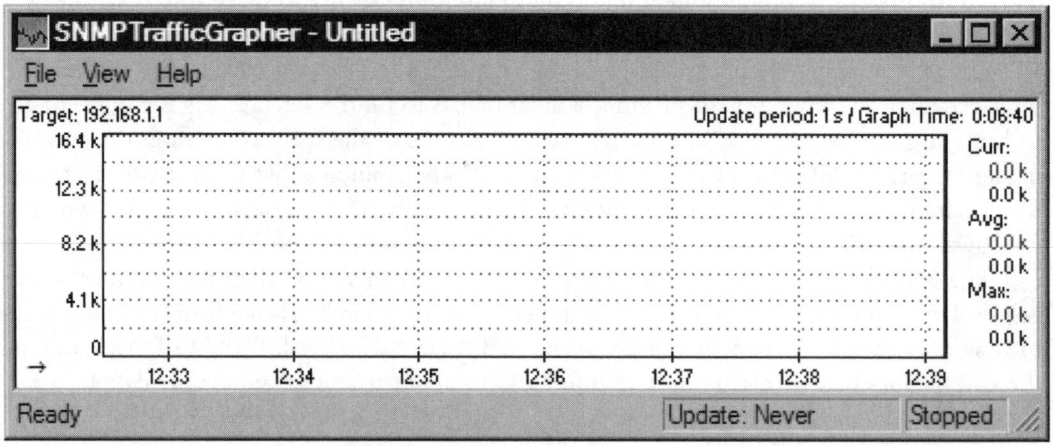

2. Go to the **View** menu and select **Settings**, or press the shortcut key, **F9**. This will display the settings window shown in Figure 7.10.

Figure 7.10 STG Settings

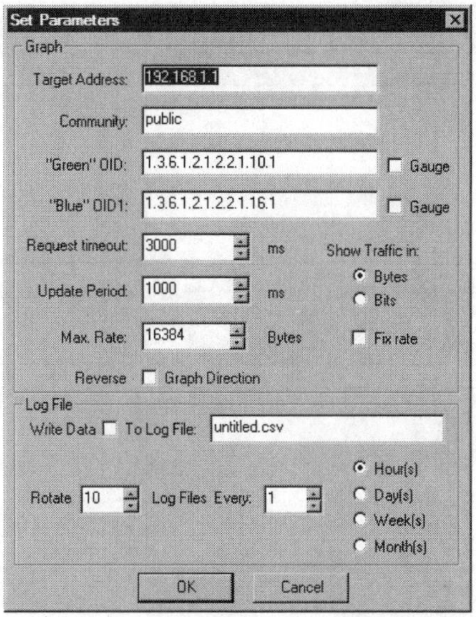

3. If you point the target address to your SNMP device, and then click **OK**, it will query the device every second for the inbound and outbound data transfer for its first interface.

How does it know to do this? The secret is in the two "Green" and "Blue" OID fields. If you examine Figure 7.7, you'll see the Blue OID setting shown in the bottom of the Getif screenshot. In this case, OID `1.3.6.1.2.1.2.2.1.16.1` is the received bytes for the m0n0wall wireless adapter interface (Figure 7.5 shows the interface list).

If we wanted to monitor the WAN port of the m0n0wall firewall, we could look up the appropriate OID in Getif and change the settings. As you'll see, only the last digit of the OID changes for each different interface. So to monitor interface 3, the WAN port, you would set the OIDs to `1.3.6.1.2.1.2.2.1.10.3` and `1.3.6.1.2.1.2.2.1.16.3` respectively.

Tips and Tricks

You can save your STG settings using the File Save menu. It will remember the window size as well as the other settings. Also, when you double-click the saved .STG file, STG will automatically restart with those settings. Put this together, and you can create a set of graphs that together provide a set of useful stats for your network. Figure 7.11 shows an example of this.

Figure 7.11 Monitoring Multiple Interfaces with STG

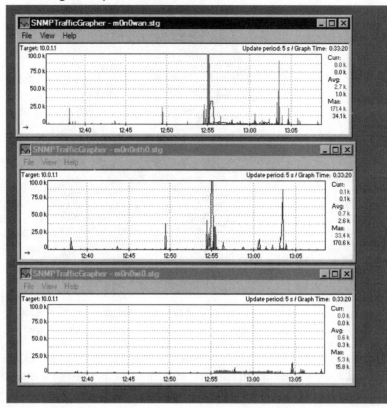

This figure shows all three m0n0wall interfaces being monitored. From top to bottom it shows the WAN interface connected directly to the Internet, then the local Ethernet LAN, and finally the wireless LAN. Notice that if you add the bottom two graphs together, you end up with the top graph. The blue line and fill colors are reversed in the top graph because the inbound LAN traffic ends up going out on the WAN interface.

Some other useful tricks:

- Provide a Max Rate value and check the **Fix Rate** box. If you choose the same scale for all the graphs, they'll be directly comparable.

- Choose a lower max rate than the interface is capable of. Even if the interface can do 6MB, a lot of the time is spent wondering if that long flat section on the graph was an outage or just a natural lull in traffic. If instead you choose a Max Rate of 100,000 (100k), you'll peg the graph occasionally, but you'll see small amounts of traffic more easily.

- Reverse the colors of your in and outbound ports so they match each other. This again makes direct comparison simpler.

- Change the **Update Period** to reflect your needs. If you're trying to debug a particular device, then you might leave it at the default of 1,000 msec (1 second). On the other hand, if you leave this on your computer all day, a 5-minute period may be better.

- Double-click the title bar of STG to enlarge the graph to full screen. This shows a lot more detail than the default small-sized graph.

- If you need to, STG can log the data and automatically rotate the logs (e.g., one per day). STG cannot view those logs however, so you may prefer another tool like MRTG if you want to capture and review historical data.

Now that you have those graphs running on your desktop computer, everyone who comes by will want to get a copy. Although they could all run the same monitoring program, that would create a lot of duplicate traffic and possibly slow down the device being monitored.

The next step in monitoring is to create a Web site that can capture and display traffic. The Cacti section will detail how to do that.

Under the Hood: How the Hack Works

Leonid Mikhailov, the author of STG, has written a clean, reliable program that does one thing well: collect and graph two OID data points. As he says, it is:

> "intended as fast aid for network administrators who need prompt access to current information about the state of network equipment."

He has intentionally modeled its appearance to be similar to the popular MRTG program. However, unlike MRTG, STG can be used quickly by copying the program to the desired machine and simply running it. By avoiding the MIB tree decoding provided by Getif, Leonid was able to keep the program small and simple.

Overall, STG is a great little utility for your toolkit.

Cacti and Comprehensive Network Graphs

A common tool for capturing network traffic is MRTG—The Multi Router Traffic Grapher. This tool periodically polls specified SNMP devices, gathers their traffic stats and builds HTML (Web) pages showing the historical usage for the past 24 hours, week, month, and year. You can download versions for both MS Windows and various Unix and Linux systems from the author's Web site at http://people.ee.ethz.ch/~oetiker/webtools/mrtg/. However, MRTG has some disadvantages since it generates new Web pages every five minutes, most of which are unused.

The authors, Tobias Oetiker and Dave Rand, have created a successor called RRDTool (http://people.ee.ethz.ch/~oetiker/webtools/rrdtool/). Unlike MRTG, this tool generates no HTML pages, but instead gathers the data into a compact format and generates sophisticated graphs on demand. The goal was to provide a base for others to build upon and that's exactly what the lead authors of Cacti, Ian Berry and Larry Adams, have provided.

Cacti is a complete HTML interface to RRDTool. Unlike MRTG which is controlled with text configuration files, Cacti presents an administration interface via a Web browser that allows configuration of everything from the polled stations to the format of the graphs. It also has a logon system that provides multiple users with varying levels of permissions (e.g., allowing them to view graphs but not alter them). Last but by no means least, it allows you to build complex graphs that combine values from multiple monitored systems. For example, you could build a composite graph showing traffic from multiple Access Points combined into one multicolored graph to show total traffic through the system and where it's coming from.

In this section you'll learn how to install Cacti on a Windows XP machine and build a basic monitoring system. The same principles apply to a Linux or Unix installation, though on Linux/Unix many of the programs will already be installed.

Preparing for the Hack

Cacti is built upon several powerful and popular free programs. Each of these needs to be downloaded and set up before installing Cacti. The steps for installing each program will be described in the following sections. Table 7.3 provides information about these programs.

Table 7.3 Cacti Installation Prerequisites

Name	Web Site	Explanation
Apache	www.apache.org	Web server
PHP	www.php.net	Scripting language used by Cacti
MySQL	www.mysql.com	Database used for storing settings
RRDTool	http://people.ee.ethz.ch/~oetiker/webtools/rrdtool	Gathers and stores data from network devices
Perl	www.activestate.com	Scripting language used by RRDTool
Cacti	www.cacti.net	Cacti program Web site

This installation of Cacti will use Apache as its Web server, thus ensuring that it will run on Windows XP Home edition (which does not include the Microsoft IIS Web server). If you have Windows XP Professional or earlier versions, you can use IIS if you prefer. See the Cacti Web site for the slight differences in installation methods.

Many of these tools come in a variety of download versions. Whenever possible, choose the Windows MSI installer option. This will be the most automated and easiest to install.

Apache

Apache comes in two major versions, 1.3.x and 2.0.x. We chose version 2.0 because it appears to be the latest stable version and likely has the best Windows installation support.

PHP

PHP is updated frequently and there are many versions available. The 5.x series is newer than 4.x, and Cacti supports both. We've chosen the latest stable 4.x release to use here. Look for "Windows Binaries" and download the "Installer" file. Also download the full Zip package since it contains some extensions that don't come in the "Installer" version.

Perl

Perl is an open source project and ActiveState maintains the best Win32 implementation. It is available for free on their Web site. Download "ActivePerl." Choose the MSI package for greater convenience.

RRDTool

Amongst the many choices to download, look for something with "win32" in it. It will most likely end with a .zip extension instead of .gz.

MySQL

Choose the "Generally Available (GA)" release for the most stability. Then find the "Windows (x86)" release. Note that this is a large download!

Cacti

There are two sets of downloads for Cacti: "cactid" and "cacti." Download the "Binaries for Windows" version of each. (Cactid is not strictly necessary, but it makes cacti run more efficiently when you have large numbers of devices to poll).

Performing the Hack

Cacti provides documentation specifically for Windows installations. It is good as a general guide, but what follows is much more detailed. Most of the work involved in setting up Cacti is actually installing all the different parts that it needs. The screen shots and details that follow assume a machine running Windows XP Home with a "clean" installation. However, Windows 2000 would likely work just as well, though probably not Windows 98 or ME. Figure 7.12 shows all the files downloaded and ready for installation.

Figure 7.12 Programs Needed for Cacti Installation

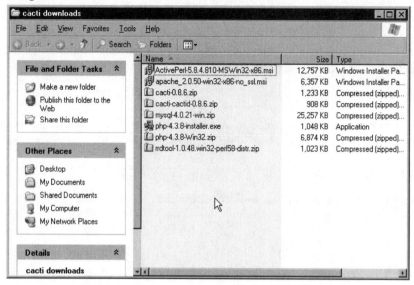

After installation, this guide continues with some basic Cacti configurations to show traffic charts from Access Points in a small network.

Installing Apache

Apache is a powerful and popular open source Web server. Cacti uses it to communicate via a standard Web browser. Some versions of Windows come with a Microsoft Web server called IIS (Internet Information Server), which you could use instead of Apache. However, as IIS is not included in Windows XP Home, these instructions assume Apache.

1. First, run the installation program, **apache_2.0.50-win32-x86-no_ssl.msi**—in this case, by double-clicking the icon (as usual).

2. When prompted for server information, as shown in Figure 7.13, you can use the defaults for the network and server name, and then enter your e-mail address. This is used in various (rarely used) places, such as the Webmaster contact.

Figure 7.13 Apache Server Information

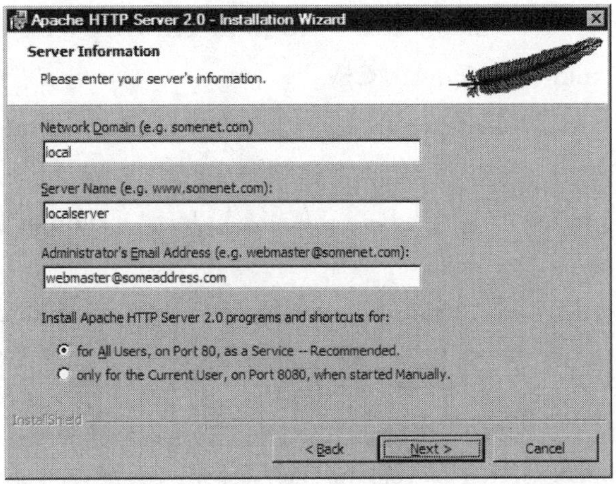

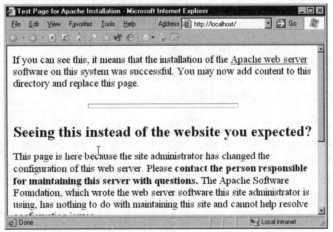

3. Continue installing with the default options: "Typical" setup type and default directory (though you could likely choose another without problems). At the end of the installation, you'll notice a new toolbar icon showing that Apache is now running. Congratulations! To confirm this, you can open a Web browser and go to the page **http://localhost/** and it should show a welcome page similar to Figure 7.14. This is the default Apache page.

Figure 7.14 Apache Default Installation Page

Now that Apache is running, we need to tweak the installation a little before continuing. Cacti does not currently support Win32 long filenames and in particular, names with spaces in them. Unfortunately, by default, Apache is installed in a path with spaces. Fortunately, there is a simple fix. You need to move the directory and all its contents and subdirectories, C:\Program Files\Apache Group\Apache2\htdocs, to the new location, c:\htdocs.

The simplest way to do this is to:

1. Open an Explorer window at **C:\Program Files\Apache Group\Apache2\htdocs**.

2. Open an Explorer window at **C:**.

3. Drag the **htdocs** folder from the first window into the second.

When you're done, the C:\ window should have a new folder called htdocs with the contents intact. Now we need to tell Apache where the new htdocs location is. To do this, perform the following steps:

1. Use the first Explorer window shown previously to navigate down to the conf directory (i.e., C:\Program Files\Apache Group\Apache2\conf) and then double-click the file **httpd.conf** to open it in a text editor (such as Notepad).

2. Search for the text **DocumentRoot** and change the line specifying the DocumentRoot from DocumentRoot "C:/Program Files/Apache Group/Apache2/htdocs" to **DocumentRoot "C:/htdocs"**.

3. Save the file.

4. Now stop and then restart Apache using the program in system tray. Open the browser again at **http://localhost** and you should see the same screen as that shown in Figure 7.14.

Despite all these steps, there are still a few programs left to install before we can see Cacti running!

Installing PHP

PHP comes packaged as an MSI (Microsoft Installer) executable. To install this program, perform the following steps:

1. Click the **php-4.3.8-installer.exe** file, or your corresponding version's filename, to start the installation process. Choose the defaults for the initial questions, such as the "Standard" type of installation and a default folder of c:\php.

   You'll be asked for an SMTP server. If at all possible, it's a good idea to provide this information. Your ISP (Internet Service Provider) will have supplied this information when you established your service. Typically, it is mail.ISPname.net (replace "ISPname.net" with your ISP). If you primarily use Web-based e-mail, then you may not have used your SMTP server before and you'll need to contact your ISP. Similarly, enter a valid e-mail address. Any messages that PHP sends will come from this address if not otherwise specified by the running script.

2. As shown in Figure 7.15, change the HTTP server to **Apache**, then click **Next** twice more and sit back while PHP is installed.

Figure 7.15 Defining the PHP Server Type

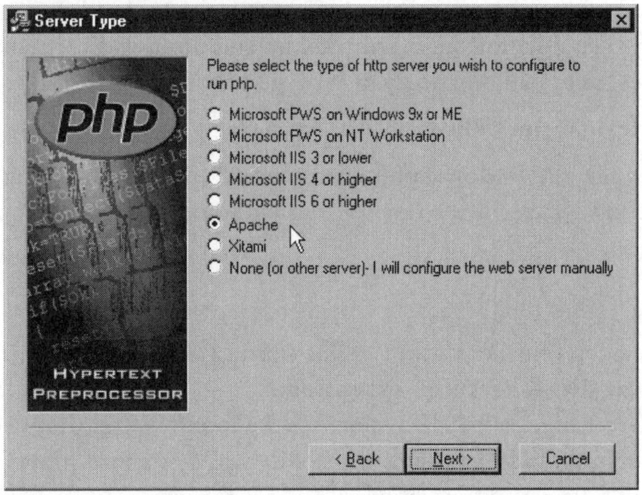

At the end, you may receive a warning that Apache has not been automatically configured, so let's jump in and tell Apache that PHP is now installed (the full details of this are in the file c:\php\install.txt).

3. Open the file **C:\Program Files\Apache Group\Apache2\conf\httpd.conf** and locate the multiple lines starting with **LoadModule** (near line 170 in this version).

4. Add the lines:

    ```
    # Added LoadModule for PHP support
    LoadModule php4_module "c:/php/sapi/php4apache2.dll"
    AddType application/x-httpd-php .php
    ```

5. Find the line starting with **DirectoryIndex** (line 326) and add index.php to the end so it now reads:

    ```
    DirectoryIndex index.html index.html.var index.php
    ```

6. Save the file.

7. Copy the file **c:\php\php4ts.dll** to your Windows system directory (typically, **c:\windows\system32** or **c:\winnt\system32** for Windows NT/2000). This completes the basic configuration of Apache and PHP, but we still need to get some extra files from the PHP Zip file distribution.

8. Open the Zip file, **php-4.3.8-Win32.zip**, and copy the following directories to the c:\php directory where PHP has been installed:

    extensions

    mibs

    sapi

A simple way to do this with Windows XP is to keep double-clicking the Zip file and its contained folders, selecting **Edit | Copy** on the folder(s) you wish to copy, and then opening a window for the **c:\php** folder and doing **Edit | Paste**. This extracts the files from the Zip file automatically as it copies them.

9. Extract the directory **dlls** from the Zip file and place its contents right into **c:\php**.

10. Now the file **c:\windows\php.ini** file (installed by the PHP installer) needs to be modified. Open it and search for the line starting with doc_root (line 421 in this version) and change it to read:

    ```
    doc_root = "C:\htdocs"
    ```

    A few lines further down is a line starting with **extension_dir**. Modify it as follows:
    extension_dir = "c:\php\extensions"
    These changes tell PHP where Apache looks to find the Web pages to show, and where PHP should find its extensions (such as MySQL support), respectively. Later versions of PHP may have already modified these lines for you.

11. Search for the line containing **php_snmp.dll** (line 574).

12. Remove the leading semicolon comment indicator for this line and the sockets.dll nearby. The two lines should read:

    ```
    extension=php_snmp.dll

    extension=php_sockets.dll
    ```

13. These extensions will be needed for Cacti. After you've made these changes, save the file. We're now done with PHP changes.

14. Stop and restart the Apache server using the system tray icon. If it can't restart for any reason, you can review the Apache error log file available from the Windows **Start -> All Programs** menu.

15. Before continuing, let's test our PHP installation. To do this, create a file called **test.php** in the directory **C:\htdocs** containing the following line:

    ```
    <? phpinfo(); ?>
    ```

    This line tells PHP to return its configuration information in the Web page. Time to see if it all works!

16. Open a browser window and type in the address **http://localhost/test.php**.

If all goes well, you should see the PHP Version information as shown in Figure 7.16. Congratulations on getting this far! This was the hardest part.

Figure 7.16 PHP Status Screen

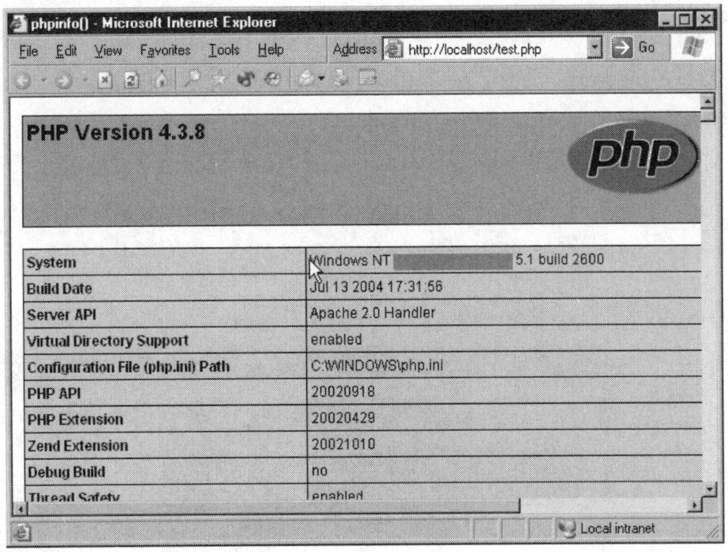

Installing Perl

Now that we have the Web server and PHP running, it's time to install Perl as a prerequisite for RRDTool. All you need to do is double-click the installation program, **ActivePerl-5.8.4.810-MSWin32-x86.msi**, and follow the prompts from the Installation Wizard. You can use all the defaults for this installation and no information is required.

No further configuration is needed!

Installing RRDTool

To Install RRDTool, perform the following:

1. Extract all the files in the Zip archive, **rrdtool-1.0.48.win32-perl58-distr.zip**, into the directory **c:\rrdtool**.

2. Open a command prompt by selecting **Start | Run**.

3. Enter **cmd** and click **OK**.

4. To complete the configuration, enter:

   cd \rrdtool

   copy src\tool_release\rrdtool.exe

   cd perl-shared

   ppm install rrds.ppd

   exit

Installing MySQL

To install MySQL, perform the following:

1. Unzip the contents of the mysql file, **mysql-4.0.21-win.zip**, and run the **setup.exe** program.

2. Use the default installation direction of **c:\mysql** and choose the **Typical** installation.

3. After it completes, run the program **c:\mysql\bin\winmysqladmin.exe**. It will start as shown in Figure 7.17 and prompt you for a username and password.

Figure 7.17 The WinMySQL Startup Screen

4. Click **Cancel** to skip the username and password since it is not necessary to store this information.

A "traffic light" icon with the green light glowing will appear in the bottom-right system tray to indicate that MySQL is running successfully.

We'll come back to MySQL to complete the Cacti configuration after installing Cacti.

Miscellaneous Settings

For PHP, Cacti, Cactid, and Apache to all work together well, some new environmental variable settings are needed. In Windows XP, these are set in the **System Properties** dialog box which is available from the **Control Panel** or by right-clicking **My Computer** and selecting **Properties**. Here's the step-by-step process:

1. Choose **Start | Control Panel | Performance and Maintenance**.

2. Choose **See basic information about your computer** (or **System Properties** if that isn't present).

3. When the **System Properties** dialogue opens, click the **Advanced** tab.

4. Near the bottom of that page, click the **Environment Variables** button.

5. In the **System Variables** section, find **Path** and click **Edit**.

6. At the end of the Variable value, add the text **;c:\php**.

7. Click **OK**.

So far, we've modified the Path environmental variable to add the PHP directory. Next, we add two new environmental variables for the SNMP support. To do this, perform the following steps:

1. Click the **New** button.

2. For **Variable Name**, enter **MIBS** and for **Variable Value** type **ALL**

3. Click **OK** to save the new variable.

4. Click the **New** button.

5. For **Variable Name**, enter **MIBSDIR** and for **Variable value** type **c:\php\mibs**.

6. Click **OK** to save the new variable.

7. Now you can close these windows by clicking their **OK** buttons.

8. To verify that these settings are correct, open a command prompt from **Start | Run** and type **cmd**.

9. In the resulting **MS-DOS** box, type **set** to see the updated Path environment variable and the new variables you entered, along with the other environmental variables that were already there.

Installing Cactid and Cacti

Cacti has two main parts, the Cacti PHP scripts that build the Web pages that show the graphs and configure the network polling, and Cactid which controls the periodic data gathering from the SNMP devices. Perform the following to install cactid:

1. Create the directory **c:\cactid** and then copy the files from the cactid Zip file, **cacti-cactid-0.8.6.zip**. In this release, there are three files: cactid.conf, cactid.exe, and cygwin1.dll. Edit the **cactid.conf** file to change the log file destination as follows (the last line):

```
LogFile           c:/cactid/cactid.log
```

Cacti itself is installed in the Apache Web files directory. Create the directory: **C:\htdocs\cacti**.

2. Unzip the contents of the archive directly into that location. Be sure not to include intermediate folder names like cacti-0.8.6. For example, in the c:\htdocs\cacti folder there

should be folders called images, include, and install as well as approximately 40 files with names like about.php, cmd.php, and so on.

3. After the files are installed, open the file **C:\htdocs\cacti\include\config.php** and confirm that it contains the following lines near the beginning:

```
$database_default = "cacti";

$database_hostname = "localhost";

$database_username = "cactiuser";

$database_password = "cactiuser";
```

The final step before running Cacti is to configure MySQL. We need to create a database to store the Cacti data and then use a configuration script to create the tables Cacti uses. The database settings listed earlier are assumed in the following instructions:

1. Open an MS-DOS command prompt by choosing **Start | Run...**

2. Enter **cmd** and click **OK**. This should open the **MS-DOS** box (a Command Prompt Shell).

3. Then enter the command: **cd c:\mysql\bin**.

4. Type:

```
mysqladmin –uroot create cacti

mysql –uroot cacti <C:\htdocs\cacti\cacti.sql

mysql –uroot
```

5. At this point, the prompt will change from the usual MS-DOS prompt of c:\mysql\bin to the mysql utility prompt. Continue with:

```
Grant all on cacti.* to cactiuser@localhost identified by 'cactiuser';

Flush privileges;

exit
```

The full sequence of commands and responses is shown in Figure 7.18.

Figure 7.18 Configuring MySQL for Cacti

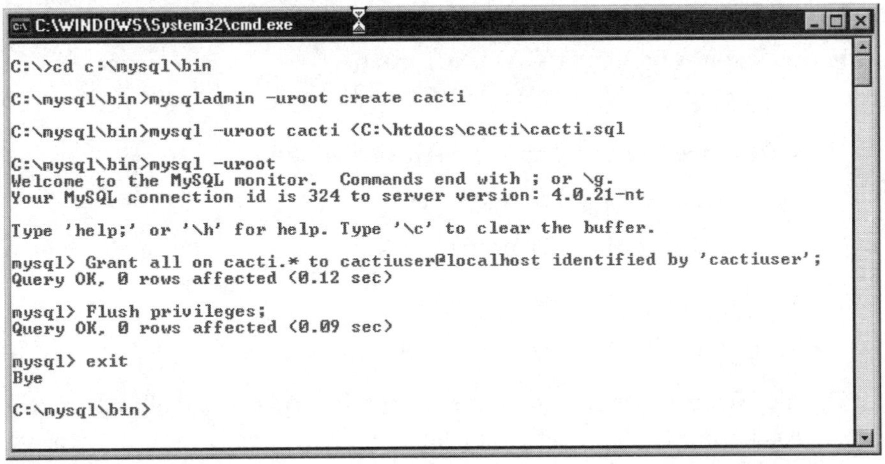

```
C:\WINDOWS\System32\cmd.exe

C:\>cd c:\mysql\bin

C:\mysql\bin>mysqladmin -uroot create cacti

C:\mysql\bin>mysql -uroot cacti <C:\htdocs\cacti\cacti.sql

C:\mysql\bin>mysql -uroot
Welcome to the MySQL monitor.  Commands end with ; or \g.
Your MySQL connection id is 324 to server version: 4.0.21-nt

Type 'help;' or '\h' for help. Type '\c' to clear the buffer.

mysql> Grant all on cacti.* to cactiuser@localhost identified by 'cactiuser';
Query OK, 0 rows affected (0.12 sec)

mysql> Flush privileges;
Query OK, 0 rows affected (0.09 sec)

mysql> exit
Bye

C:\mysql\bin>
```

WARNING: SECURITY CONCERN

By default, the MySQL server does not include a root password. This simplifies configuration, but it leaves your database server wide open to access from the outside. If you expect your server to be accessed by others, you should at minimum add a mysql password and preferably close the appropriate mysql TCP/IP ports in your firewall.

The final part of the setup is to create a task that starts every five minutes to trigger the polling of the SNMP devices.

To do so, perform the following steps:

1. Go to **Start | Control Panel**, then click **Performance and Maintenance**.

2. Click **Scheduled Tasks** (if you have *Classic View* Control Panel, you can go straight here).

3. Double-click **Add Scheduled Task** to start the **New Task Wizard**.

4. Click the **Browse** button and select the file **c:\php\php.exe**.

5. On the next wizard page, optionally change the task name to **Cacti** and choose **Daily**.

6. Click **Next** to get past the time and date settings.

7. Enter your logon password. If you don't have a logon password, you'll have to create one via the **Control Panel | User Accounts**, because task manager won't run your task without it. Click **Next**. (Fortunately, there's a TweakUI tool to automatically log you on at XP startup if you're the only user and want to avoid the inconvenience.)

8. On the last page, check the **Open advanced properties...** option and then click **Finish**.

9. In the **Advanced** properties box that follows, select **Run**, and enter **C:\PHP\php.exe C:\cacti\poller.php**.

10. In the **Start In** box, enter **C:\htdocs\cacti**.

11. Click the **Schedule** tab.

12. Click **Advanced** to bring up more scheduling options.

13. Check the **Repeat Task** box and specify **5** minutes.

14. Change the **Duration** to **24 hours**.

15. Click **OK**.

Now you're ready to start Cacti. To do so, follow these steps:

1. Open a Web browser and go to the Cacti URL: **http://localhost/cacti**.

    After a short delay, Cacti will automatically redirect you to its installation page, as shown in Figure 7.19. Congratulations for getting this far. You're almost finished!

Figure 7.19 The Cacti Web Site Installation Screen

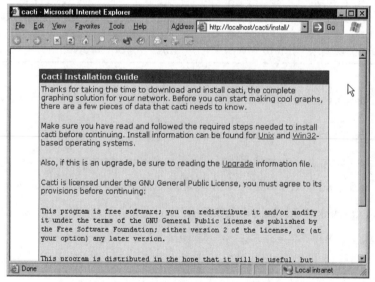

2. Scroll to the bottom of this introductory page and click the **Next>>** link.

3. On the following page, leave the drop-down choice at New Install, confirm that the settings match what we've previously specified, and again click **Next>>**.

4. The next page will confirm the location of RRDTool, the PHP installation, and the Cacti Log File Path. If everything is working correctly, each will have the text **[FOUND]** above them.

5. Click **Finish** and you're done! If all has gone well, you'll now see the Cacti login page shown in Figure 7.20.

Figure 7.20 The Cacti Login Screen

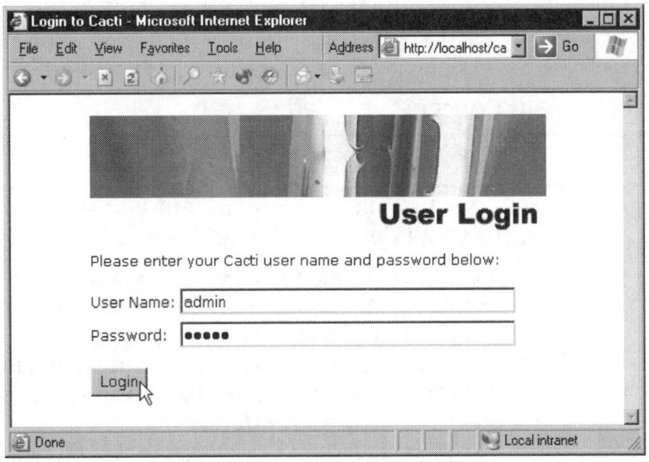

6. Log into Cacti with the initial User Name of **admin** and Password **admin**. You'll immediately be asked to enter a new password. When that is finished, the main Cacti screen will appear, as shown in Figure 7.21.

Figure 7.21 The Main Cacti Screen

You're now ready to start gathering and graphing statistics!

Graphing Data in Cacti

Now that you have Cacti running, it's time to add the devices you want to monitor. A quick overview of Cacti will help before jumping in. Cacti has three main concepts. Table 7.4 describes them.

Table 7.4 The Three Main Concepts of Cacti

Concept	Description
Device	An actual piece of equipment you wish to monitor. You can think of this as a device addressable with an IP. Also called a Host by Cacti.
Data Source	Individual items (or pairs) of data gathered from devices. For example, the traffic statistics for a single interface on a device is a data source. Thus, a typical AP device will have at least two data sources: traffic through its Ethernet interface and traffic through its wireless interface. Other data sources for an AP might include the error data for each interface.
Graph	A picture showing one or more data sources, possibly combined in various ways (e.g., summed, totaled, averaged).

You build a graph for a new device using a three-step process:

1. Describe the device.

2. Choose which data you wish to collect from the device.

3. Build the graphs based on the data you're collecting.

This three-level approach seems cumbersome at first, but in practice it's very powerful. For example, if a device moves to another IP address, the only change you need to make is to the device settings. Everything else remains unchanged. Or, if you wish to create a combined graph showing traffic from multiple devices, you can choose from the various data sources available and then combine them arithmetically to create just the graph you want—and still have all the previous graphs. And, because these all draw from the same data, the data is only collected and stored once, which saves computing and storage resources.

In addition, Cacti has templates to use with each of these concepts which pre-populate many common values. You can safely ignore these until you've mastered basic Cacti features.

The following set of screen shots shows the setup for monitoring a device running a m0n0wall firewall AP, with a WAN (Internet) connection, a LAN connection, and a 802.11b radio. The configuration is similar for other SNMP devices. Simply perform the following steps:

1. Delete the existing **localhost** entry. This is erroneously configured for a Linux box and won't work under Windows XP.

2. Check the box on the far right of the Localhost line, ensure the action says **Delete** and then click **Go**, as shown in Figure 7.22.

Figure 7.22 Delete the Localhost Entry

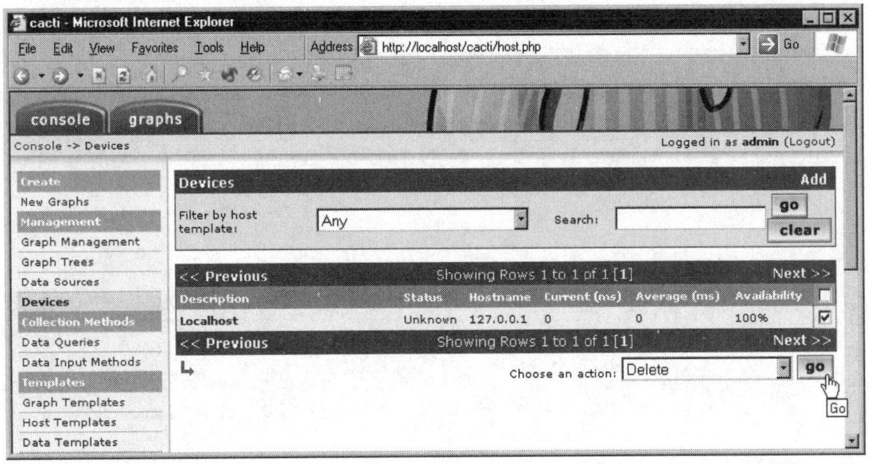

3. Click **Yes** on the subsequent confirmation screen, using the default choice of removing all graphs as well as the device. This will return you to the **Devices** screen.

4. In the top-right corner, click the word **Add**, right above the **Go** button. A **new device** form will appear, as shown in Figure 7.23.

Figure 7.23 Adding a New Device

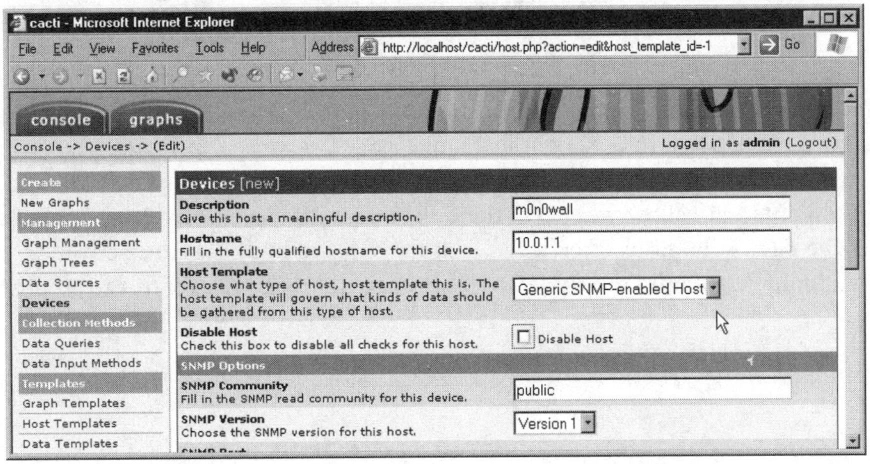

The first three boxes are the important ones to fill in. The *description* should be your name for the device being monitored. This will appear on graphs and in most places within Cacti. You might base it on the type of device (e.g., m0n0wall), or, if you have multiple similar devices, on the location of the device (Downtown m0n0wall). The *Hostname* is the Internet address for the device. In a large network, this is mostly likely registered in DNS somewhere, but for a home LAN a simple IP will work just as effectively. In the screenshot shown, 10.0.1.1 is the internal address of a system running m0n0wall (described in Chapters 3 and 6). The *Host Template* describes the type of device. The "Generic SNMP-enabled Host" shown is a good choice for most SNMP devices if you only wish to monitor traffic. (Note that the template pre-fills certain choices but doesn't lock you into that choice. You can always add more settings later.) If your information is correct, the screen shown in Figure 7.24 will appear. Note the *SNMP Information* summary provided under the main heading. This confirms that your host information was correct and that Cacti was able to communicate directly with the device.

Figure 7.24 The Device Was Created Successfully

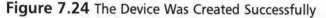

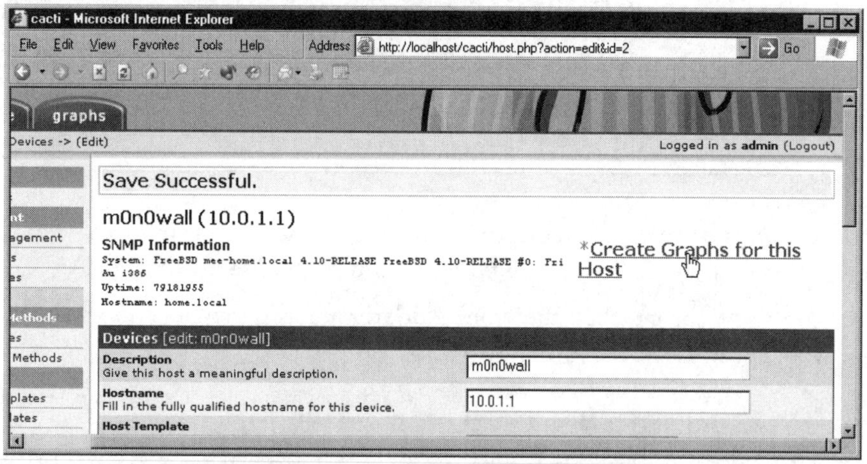

Now we have a device and can create data sources and graphs for it. Cacti conveniently combines this into one step using information from the host (device) template you selected. Click the **Create Graphs for this Host** link, as shown in Figure 7.24. The screen in Figure 7.25 will appear. In the section marked **Data Query**, click the checkboxes for the interfaces you wish to graph. Each line will be highlighted in yellow whenever the box is clicked. Note how Cacti created this list by querying the device for all its interfaces. Click the **Create** button at the bottom to complete the selection. Figure 7.26 shows the subsequent screen with confirmation of the graphs that were created.

Figure 7.25 Creating Graphs for Device

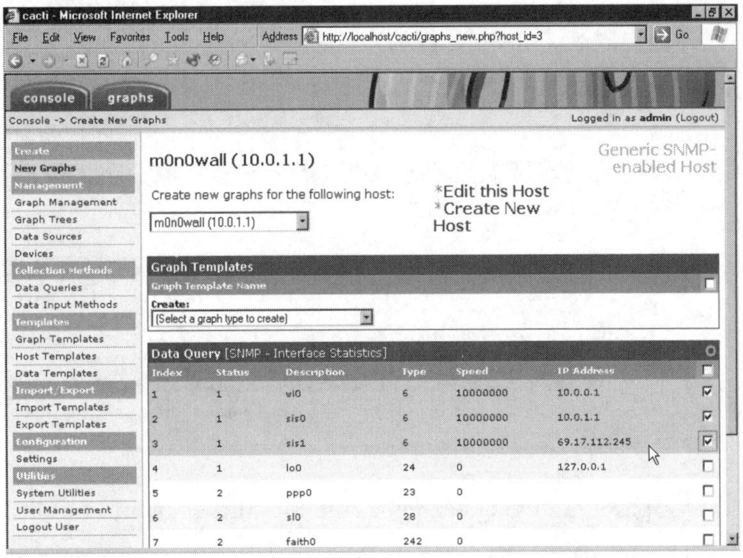

Figure 7.26 Graphs Created Successfully

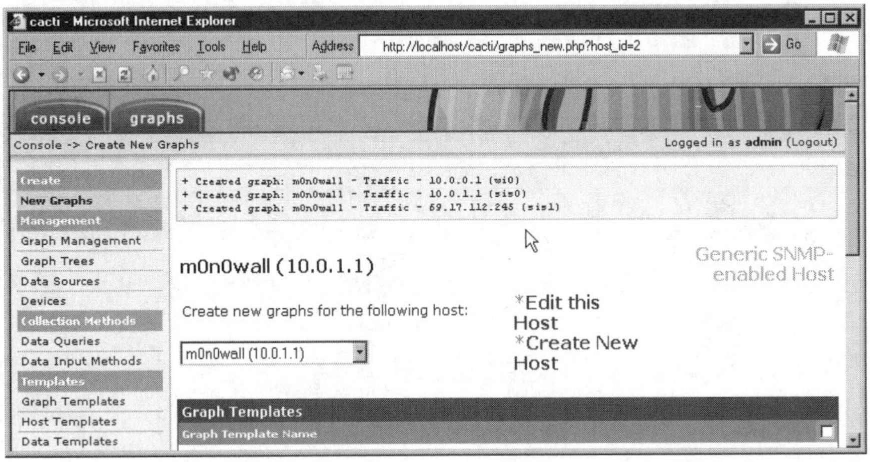

Now the device is defined, the data sources are specified, and the graphs, now defined, will be created as required. The only thing remaining is to display them!

1. Click the **Graph Management** link in the left-hand column to display the screen shown in Figure 7.27. Cacti arranges its graphs in *trees,* which are organized by you. This tree structure is handy for showing summary information at the root of the tree, followed by increasingly more detailed information in its branches. So far, there are only three graphs, so they can all go at the tree root. Figure 7.27 shows how to do this.

Figure 7.27 Graph Management for a Default Tree

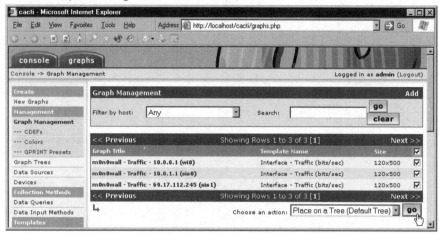

2.   Select all the graphs using the checkbox at the top of the column on the right.

3.   Choose the action that says **Place on a Tree (default tree)** and click **go**.

On the next screen is a confirmation and also a tree location selection. In this case, leave the default *[root]* and click **yes** to confirm your choice. This is shown in Figure 7.28.

Figure 7.28 Confirming the Tree Placement of Graphs

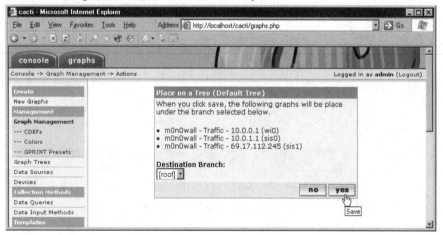

Now that you have the default graph display tree configured, you can click the large **graphs** tab on the top left of the screen. After an hour or so, the graphs will accumulate enough data to appear as shown in Figure 7.29. This shows the first of three graphs available on this page. If you click a graph, it will show another page of historical graphs displaying the weekly, monthly, and yearly historical usage of that same data.

Figure 7.29 Cacti Graphs!

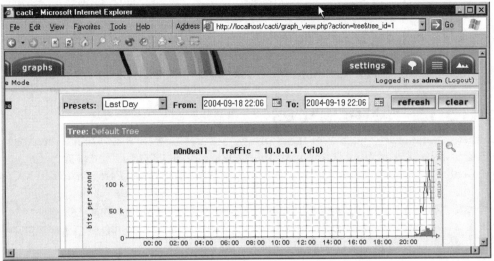

Graphs of an individual interface are interesting and useful to review, but as the number of devices and interfaces in your network increases, it's more convenient to view a small number of graphs to get the big picture and only view individual interface graphs as needed (e.g., debugging a network problem). Cacti has a powerful set of tools to help you do this. See the Cacti documentation at their Web site, www.cacti.net, for more information.

Under the Hood: How the Hack Works

The many programs you have installed have turned your computer into a powerful server that not only provides Web pages, but also polls whatever devices you have specified at regular intervals.

Every five minutes, the Windows Task Scheduler tells PHP to run a page directly (i.e., without the Apache Web server being involved). This in turn starts a polling program which can either poll the devices directly via SNMP, call a user-supplied script to retrieve specific information, or call the separate cactid executable to poll multiple SNMP devices more quickly.

When it has gathered this data, the polling program passes it to RRDTool. Its author, Tobi Oetiker, explains its function well:

> RRD is the Acronym for Round Robin Database. RRD is a system to store and display time-series data (e.g., network bandwidth, machine-room temperature, server load average). It stores the data in a very compact way that will not expand over time, and it presents useful graphs by processing the data to enforce a certain data density.

If you never look at the Web pages showing the statistics, this is all that happens. The data is accumulated, but no graphs are generated or displayed. If a browser requests a page from Apache which contains a graph, then the Cacti PHP code is triggered. This then calls RRDTool with all the parameters needed to make the specific graph requested. From this, it generates a Web page containing the graph you can view. This "build on demand" approach uses resources very efficiently. (Cacti also has an option to generate graphs at regular intervals, which can be useful when built into static Web pages.)

Cacti uses the MySQL database to store all the settings it receives via the Web interface. All the device information, requested graphs, and templates are stored in the database. Using a database allows Cacti to easily devise the appropriate graph generating command and polling commands.

Additional References

There are many more tools available for monitoring systems. Google has two useful directory pages at http://directory.google.com/Top/Computers/Internet/Protocols/SNMP/ and at http://directory.google.com/Top/Computers/Software/Networking/Network_Performance/RMON_and_SNMP/.

The next level of monitoring tool provides notification (via e-mail or pager) of unusual network events, such as a server that no longer responds, or monitored values moving outside of specified limits. Some good starting points include the following:

- Nagios: www.nagios.org/

- Big Brother: http://bb4.org/

- http://directory.google.com/Top/Computers/Open_Source/Software/Internet/Monitoring/

## Chapter 8

Low-Cost Commercial Options

Topics in this Chapter:

- Sputnik
- Sveasoft
- MikroTik

Introduction

Community wireless networks can be created using a variety of funding scenarios. Sometimes, a project will have a sponsor who will pay for hardware costs. Other times, a project has no funding source whatsoever and operates on a shoestring budget. This book outlines many open-source and free options that are available to help deploy a wireless network. However, a "free" solution is not necessarily always the "best" solution, as every installation is unique and no one solution is best for all deployments. However, having a project sponsor does provide some flexibility and more options for hardware and software. While open source does have many advantages (such as being free!), one of the nice aspects of using a commercial solution is that professional support is available. If you run into problems or have questions, you can usually get help right away. In this chapter, we review three low-cost commercial options.

Sputnik

Community wireless networks shouldn't become a victim of their own success. All too often, groups plunge into network deployment projects without any vision for large-scale network management. Setting up one access point (AP) is easy. However, the task of trying to keep track of dozens of APs, monitoring users, upgrading firmware, and keeping the network operational is an overwhelming task that many people underestimate.

Successful models for building community wireless networks always revolve around the persistent question of, "What will this network look like in 100 nodes?" If you don't plan for growth now, you can be sure that one day, you will pay the price in the form of an unreliable network, unhappy users, and unacceptably high levels of unscheduled downtime.

With Sputnik, you can deploy and manage a large-scale Wi-Fi network with ease! The Sputnik platform provides easy provisioning, network- and user-level management, real-time monitoring, and remote upgrades. Sputnik is a stroke of genius for community wireless networks that are serious about large-scale growth. Let's see how it works.

Sputnik Access Points

Sputnik uses special APs that incorporate the "Sputnik Agent," which is a special firmware written specifically for that device. At the time of this writing, Sputnik has agents for two APs, the AP-120 and the AP-160. However, additional Sputnik Agent ports are currently in development. The AP-120 is an inexpensive, entry-level 802.11b device designed for indoor use. The AP-160 adds 802.11g capabilities along with external antenna support (RP-SMA connector) and a four-port switch for adding additional devices. Figure 8.1 shows the AP-120, and Figure 8.2 shows the AP-160.

Figure 8.1 The Sputnik AP-120

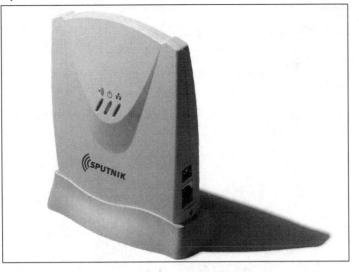

Figure 8.2 The Sputnik AP-160

While many locations already have wireless hardware in place, you can simply integrate Sputnik into an existing deployment by using an AP-160 and connecting the legacy APs to one of the switched ports. Even more efficient is to use the WDS functionality built into both the AP-120 and AP-160.

While the AP-120 and AP-160 are designed for indoor use, Sputnik also offers an AP-200 specifi- cally designed for the outdoors. This rugged 802.11b device features a 200mW radio, along with external antenna support (N connector) and optional Power over Ethernet (PoE). The AP-160 and

AP-200 make a great combination. You can connect the AP-160 to your DSL or cable modem and then run cat5 to the AP-200 or use WDS to let the devices communicate wirelessly! Figure 8.3 shows the AP-200.

Figure 8.3 The Sputnik AP-200

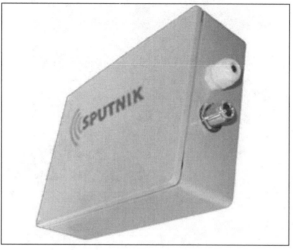

Sputnik Control Center

Each Sputnik-controlled AP (running the "Sputnik Agent") communicates over the Internet with the Sputnik Control Center (SCC). Using the SCC, you can monitor and manage the entire network remotely from anywhere on the Internet. The intuitive and easy-to-use browser-based management interface gives you access to AP configuration options and real-time reporting functionality. You can modify or monitor any aspect of wireless operations, such as changing the Service Set ID (SSID) or channel for any AP! From the browser-based management interface, you can even ping, reboot, or upgrade firmware for any AP… all remotely! Figure 8.4 shows screen shots of the SCC interface.

Figure 8.4 SCC Interface

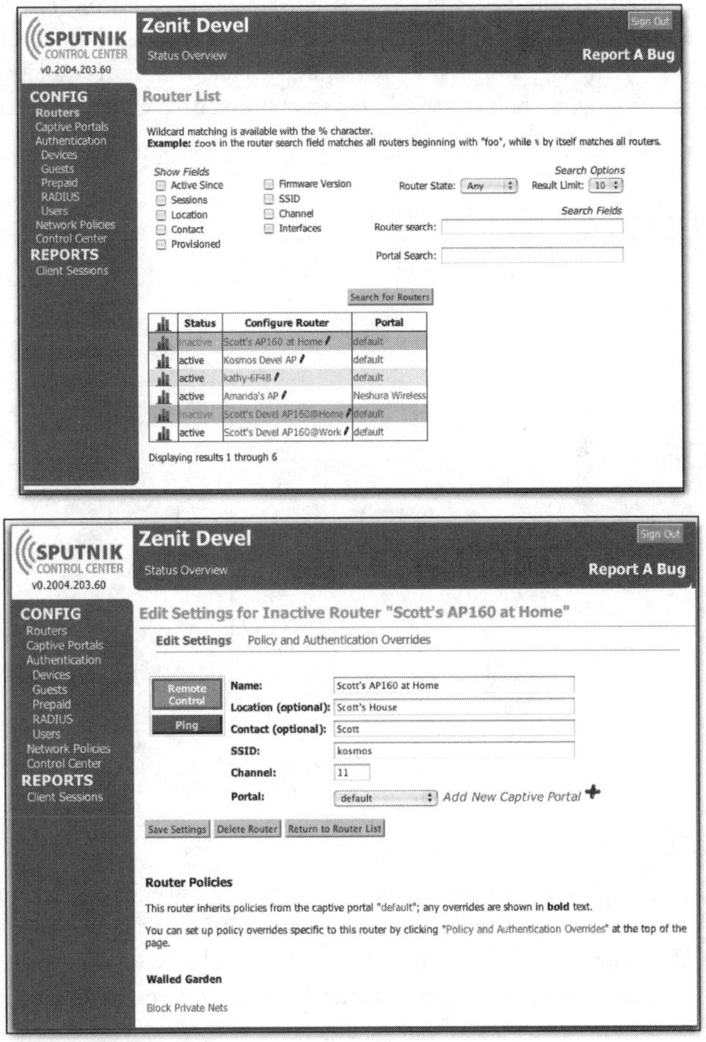

System requirements for the SCC:

- Red Hat Linux Enterprise Edition 3.0, Fedora Core 1, or White Box Enterprise Linux
- Intel Pentium II–class processor
- 64MB RAM
- 2GB hard drive
- Ethernet network interface card (NIC)
- Keyboard, monitor, mouse (PC-standard)

Note that if you are unable or choose not to run your own SCC, there are other options available for you. Sputnik offers a hosted solution, called SputnikNet. Using SputnikNet, you can purchase a Sputnik-enabled AP and then configure it to operate on a SputnikNet server instead of using your own SCC. This is a convenient solution if you don't have access to a high-availability data center, lack Linux expertise in your group, or prefer to leave server maintenance tasks to somebody else. Figure 8.5 shows a typical Sputnik deployment architecture.

Figure 8.5 Sputnik Deployment Architecture

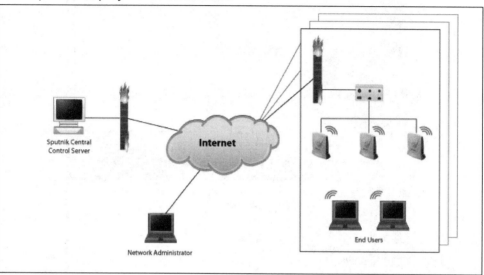

Sputnik Features

With Sputnik, it's easy to deploy and provision new APs. Because everything is centrally managed, you can enjoy a system with tremendous flexibility and scalability. Two of the most exciting features of the Sputnik platform are the Captive Portal and the Pre-Paid Module. Note that Sputnik also offers RADIUS integration support, moving Sputnik towards the enterprise tier of products.

Captive Portal

Using a Captive Portal, property owners can protect themselves from legal liability by providing a Terms of Service (ToS) page that their users must agree to before gaining access to the network. The way in which the Captive Portal works is that the AP "redirects" any Web request to a specific page, until the user clicks **I Agree** to the ToS. Therefore, when you open a Web browser, you will see the Captive Portal page first—regardless of what URL your browser initially requests.

Using the SCC, you can define a captive portal by simply entering the text or HTML directly into the management interface. Figure 8.6 shows the Captive Portal Properties page for the default portal.

Figure 8.6 Captive Portal Properties

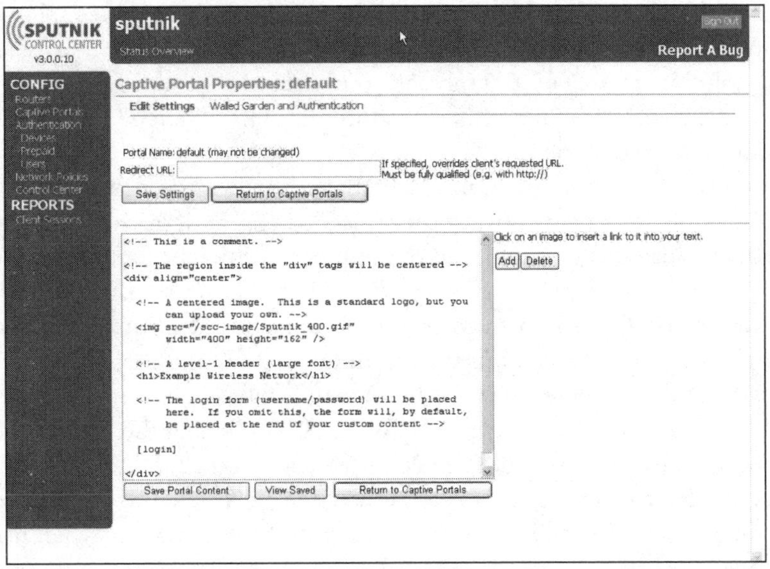

Next, you can select any router in the Sputnik cloud and assign any Captive Portal to that router. In this manner, each AP can have its own unique Captive Portal screen, or the same Captive Portal screen. You could even define several different Captive Portals and assign them to different APs at will—you are only limited by your imagination. Updating the Captive Portal is easy. With a few clicks, you can modify all of your APs with a message of the day, or an urgent notice about upcoming maintenance.

Sputnik allows you to force users to authenticate and establish accounts using a built-in database (or with optional hooks into a RADIUS server). Alternatively, you can establish guest access so that users are not required to authenticate, but instead, merely have to click **OK** to accept the ToS and gain access to network resources. The Captive Portal also has a "Walled Garden" feature so that you can exempt certain destination hostnames or IP addresses from the Captive Portal. In this way, you restrict users from accessing the Internet before they authenticate, *except* for certain Web sites, such as your own homepage or other sites that you might want users to be able to see before they log in. In addition, Sputnik supports MAC address based "whitelists" so you can permit certain devices to always be allowed access. This is useful for supporting "browserless" devices, such as Wi-Fi phones and scientific equipment.

Pre-Paid Module

Using the Pre-Paid Module allows you to generate "one-time use accounts" that can be customized for each location with a predetermined amount of access time. In other words, if a coffee shop owner is worried about users "camping out," he can issue unique passwords that limit Internet access to a specific amount of time. Then, he can print up cards and hand them out to customers using any method he chooses. Using the Pre-Paid Module in a community wireless context, the Wi-Fi access becomes "Free with purchase." This is a fair way to ensure that the coffee shop owner's generosity is not taken advantage of by users who take up space but fail to patronize the establishment. Instead of resorting to sneaky tactics like eliminating power outlets and leaving laptop users with empty batteries, the Sputnik solution allows you to embrace your customers, encourage Wi-Fi use, and at the same time protect the business. It's kind of like a bathroom with a "token" based door lock to limit transient access, vandalism, and abuse.

To create the Pre-Paid accounts, you can either use the built-in generator, or upload a .csv file. With the generator, you enter a name for the particular "batch," a username prefix, a starting suffix number, the number of accounts to create, and the number of minutes for each account. The number of minutes can be configured using one of two settings: **Time is one continuous block from first use, connected or not** or **Time is discontinuous blocks spent connected to the network**. Let's say, for example, that you set the number of minutes at 60. Using these choices, you can specify if the time expires exactly 60 minutes after the first logon, or if the customer can use 30 minutes today, 10 minutes tomorrow, and 20 minutes next week. After clicking the **Execute** button, Sputnik creates a table showing UserID, Password, Type, Minutes, and Status. You can click a link to download the accounts as a .csv file, which is useful for doing data merges in a Word document and creating custom cards for the location. Figure 8.7 shows the output of the generator when using a username prefix of "test," a Starting suffix number of "111," five accounts to create, 60 minutes each, and set to "Time is one continuous block from first use, connected or not."

Figure 8.7 Output of the Pre-Paid Module Generator

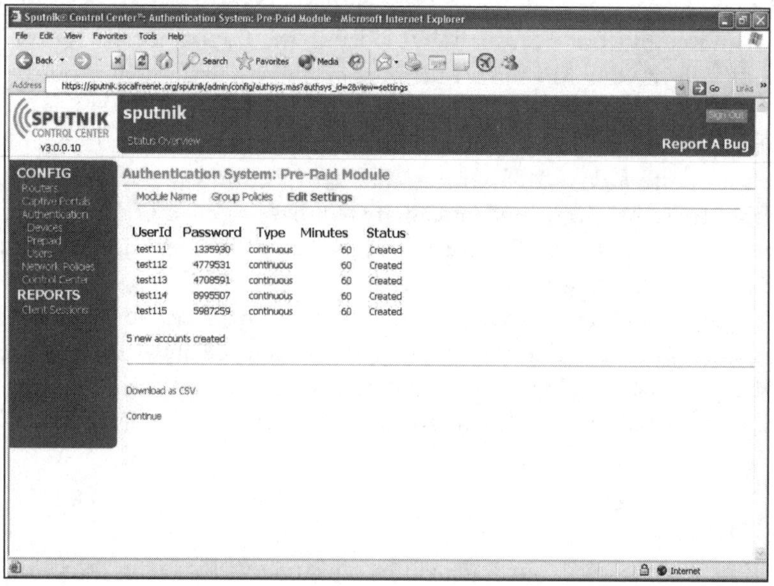

A Sputnik Revolution

With Sputnik, you can rapidly deploy large-scale wireless networks with ease. The centralized management functionality of the SCC means that you can grow your footprint and still be able to manage all of the APs in a single browser-based interface. If you prefer not to require user accounts, you can configure Sputnik to treat everybody as a guest. Alternatively, you can require the creation of user accounts and then track bandwidth use by individual user. Sputnik also includes the ability to create groups and then apply unique network policies to those groups. For example, you can allow/deny access based on Protocol, Hostname, IP, Network/Netmask, Transmission Control Protocol (TCP) or User Datagram Protocol (UDP) port, or Media Access Control (MAC) address.

Perhaps the best feature of Sputnik is its amazingly low cost. Sputnik is priced at a fraction of the cost of other products on the market with similar functionality. For current pricing and more information on the Sputnik management platform, visit their Web site at www.sputnik.com.

Sveasoft

While the Sputnik solution offers scalable hotspot management functionality, the next two solutions we will review (Sveasoft and Microtik) are better categorized as "smart routers" with wireless added. As discussed in Chapter 4, Sveasoft offers firmware upgrades for the Linksys WRT54g and WRT54gs. Interestingly, Sveasoft releases "public" versions, which are available for free, and "pre-release" versions, which are only available to subscribers. The "public" version is reviewed in Chapter 4. For $20/year, subscribers can download the latest versions of the firmware, and gain access to the message boards (technical support) at the Sveasoft Web site. For more information on the Sveasoft licensing model, visit www.sveasoft.com/modules/phpBB2/viewtopic.php?t=4277.

As of the time of this writing, the most recent pre-release version of Sveasoft is Alchemy-pre5.3. According to the Sveasoft Web site, the following features are available in this version:

Feature Additions to Alchemy-pre5.3

Client/Bridge mode for multiple clients

Adhoc mode

WDS/Repeater mode

WPA over WDS links

Web based wireless statistics

SNMP

Remote NTOP statistics

Captive portal

Extensive firewall enhancements

- track/block P2P, VoIP, IM, many other services by protocol

- 99% of available iptables filters included

VPN client and server (PPTP in all versions, IPsec as a compile-time option)

DHCP static IP assignment to specific MAC addresses

Wake-On-LAN functions with scheduling

OSPF routing

OSPF load balancing

Multi-level bandwidth management (Premium, Express, Standard, Bulk)

- manage P2P, VoIP, IM connections

- also by ports, IP addresses, and/or MAC addresses

Hardware QoS for the 4 LAN ports

Power boost to 251 mw

Afterburner Support (GS models)

SSH server and client

Telnet

Startup, firewall, and shutdown scripts

Safe backup and restore

VLAN support

Clone Wireless MAC

Reset router on upgrade

Dynamic download interface for router customization (coming)

Load balancing across multiple routers (coming)

Instant Hotspot/Captive portal with Paypal billing (coming)

External Program Support

Wallwatcher

Firewall Builder

MRTG

Cactus

PRTG

Ntop

According to the www.linksysinfo.org Web site, if you were running the Alchemy-pre5.2.3 version, the new Alchemy-pre5.3 version is primarily a bug fix release and includes the following additions:

Alchemy-pre5.3 additions from Alchemy-pre5.2.3

Dropbear V0.44test3 including ssh client

Busybox V1.0-RC3

Linksys source code and drivers V2.04.4

Bugfixes for site survey in Status->Wireless

Bugfixes Backup & Restore

Chillispot 0.96

Fixed WPA for WDS

Added ipp2p filter for P2P blocking and QoS

Fixed Access Restrictions bug

Many many small bugfixes and tweaks

MikroTik

Based in Latvia, MikroTik has been developing commercial wireless routers since 1995. While relatively unknown in the United States, MikroTik has enjoyed growing popularity in many countries around the world, including Sri Lanka, Ghana, Moldova, Albania, Estonia, Lithuania, Denmark, and more. MikroTik offers both a hardware and software platform. The hardware platform, called a RouterBOARD, is an all-in-one hardware appliance. MikroTik makes both indoor and outdoor versions of the RouterBOARD. Figure 8.8 shows the RouterBOARD 230 indoor kit, and Figure 8.9 shows the RouterBOARD 230 outdoor kit.

Figure 8.8 RouterBOARD 230 Indoor Kit

Figure 8.9 RouterBOARD 230 Outdoor Kit

Alternatively, MikroTik offers their RouterOS software as a stand-alone product that you can use with your own hardware, such as a standard PC or a Soekris device. At the time of this writing, the most current version of MikroTik is 2.8. According to the MikroTik Web site, RouterOS features include:

- Advanced wireless performance
- Even more powerful QoS control
- P2P traffic filtering
- High availability with VRRP
- Advanced Quality of Service control
- Stateful firewall, tunnels and IPsec
- STP bridging with filtering capabilities

- Super high speed 802.11a/b/g wireless with WEP
- WDS and Virtual AP features
- HotSpot for Plug-and-Play access
- RIP, OSPF, BGP routing protocols
- Gigabit Ethernet ready
- V.35, X.21, T1/E1 synchronous support
- async PPP with RADIUS AAA
- IP Telephony
- remote winbox GUI admin
- telnet/ssh/serial console admin
- real-time configuration and monitoring

Perhaps the most unique feature of RouterOS is the P2P (Peer-To-Peer) options offered by the system. This feature allows you to "shape" P2P traffic to either block it completely or to ensure that it doesn't overwhelm other traffic in the system. MikroTik constantly updates their P2P support to track the new P2P protocols appearing over time.

If you're looking for super easy configuration via a Web-based interface, RouterOS is not the answer. Their preferred programming method is via a command prompt where you enter commands like:

```
[admin@MikroTik] /ip firewall src-nat add src-address=192.168.0.0/24 out-
interface=Public action=masquerade
```

to turn on NAT to share a single public Internet address amongst multiple computers. The complexity reflects the many, many features that RouterOS offers. The functionality is there, but it's harder to configure.

There is also a GUI interface that runs on Windows machines called "Winbox" that provides a convenient way to review and tweak settings, as well as show network trace activity and traffic graphs. Figure 8.10 shows WinBox in action:

Figure 8.10 Winbox GUI interface

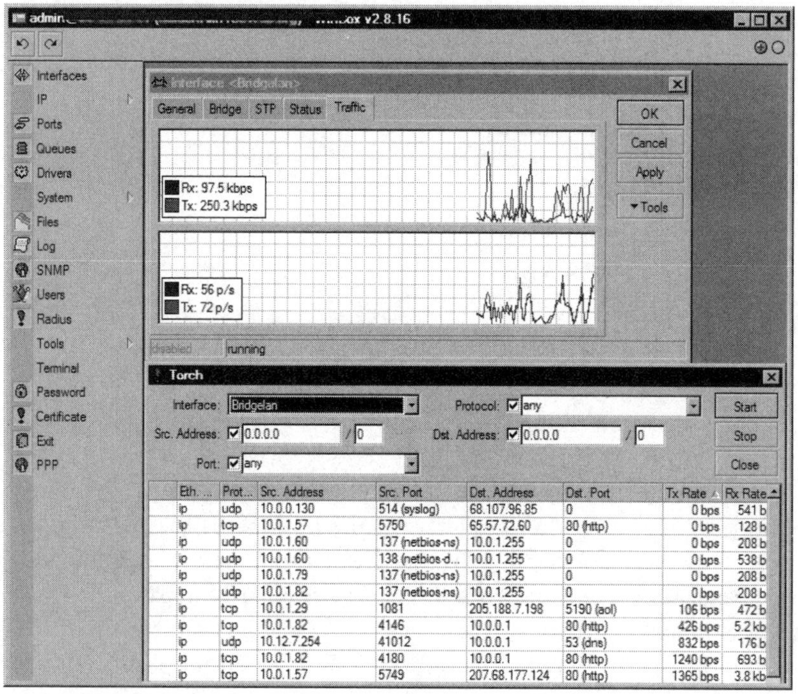

There is a simple backup and upgrade procedure that makes it easy to save your settings or upgrade to a new release if needed to add a new feature or fix a specific bug. If you don't want to roll your own BSD or Linux based solution and tweak your kernel, MikroTik's RouterOS will give you the same power and flexibility, without the hassles of building an operating system.

Mikrotik's website at www.mikrotik.com provides a wealth of information and a list of dealers around the world. There are several USA dealers listed, including www.wisp-router.com, who provide great hardware and software support.

Summary

In this chapter, we reviewed low-cost commercial options such as Sputnik, Sveasoft, and MikroTik. While many community wireless projects use open-source solutions to save costs, commercial options are available that offer excellent functionality at low price points.

Perhaps the best example of a low-cost commercial solution is Sputnik, which offers a convenient and centrally managed architecture, along with simple provisioning and deployment. The Sputnik interface is intuitive and easy to use. It packs a huge number of features into an incredibly low-priced product, and is an excellent choice for building and deploying low-cost community wireless networks.

Sveasoft offers a firmware upgrade for a WRT54g. Older versions of the firmware are available for free, while the newest "pre-release" versions are available only to subscribers who must pay a $20/year subscription fee. While the feature list for the Sveasoft firmware is impressive, you are limited to a single hardware platform, which may prove problematic for some deployment scenarios. Regardless, it is an excellent example of a quality firmware and is an enormous improvement over the stock Linksys code.

MikroTik's RouterOS platform has been used extensively overseas by the Wireless ISP (WISP) community, but does not yet enjoy wide deployment in the United States. The RouterOS has a long list of features and can operate on a stand-alone PC, Soekris hardware, or a RouterBOARD appliance that you can purchase directly from MikroTik.

## Chapter 9

Mesh Networking

Topics in this Chapter:

- Networking Magic

- Real-World Examples

- Other Resources on the Web

Introduction

Now that you have dragged in a backhaul to the Internet, programmed the router, provisioned wiring to supply power and data or supplied power with a solar panel, selected an appropriate antenna, climbed a rooftop to align it, placed an Access Point (AP) in a geographically appropriate location protected from the elements, selected wireless channels and numerous related settings, configured the wireless client, climbed the rooftop in the remote location, and repeated this RF dance for your next hop, you might be wondering if there is a better way.

Our purpose here is the exploration of alternative network topologies that address the unique deployment requirements of wireless or Radio Frequency (RF) connectivity. These alternative methods often include hardware, but the core difference lies in the software or firmware and the algorithms that control routing. Hardware differences are implemented to control phenomena inherent in RF networks that have a negative impact on the topology of choice. Operating systems can exploit the best features of a given design at the expense of standard operating procedure. Most notable examples are designs that do not follow 802.11 protocols, such as dedicated bridges, IEEE 802.15.4, IEEE 802.16 (WiMax), and backhaul links. In some cases, the deviation from published standards is purely a marketing strategy to differentiate the product from other commercial designs despite claims to the contrary. On the flip side, public domain or GNU implementations must remain vendor neutral and free of infringement but often lack the smooth integration of the full suite of options offered by enterprise equipment. This makes for a very lively mix of RF solutions.

This treatment is by no means inclusive. It is meant to inspire and provide the groundwork for additional exploration. For this chapter, we'll review the network topology and how the progress made in RF relates to cheaper and easier to deploy systems used to create communities. You will learn how to backhaul and distribute access to the Internet using Mesh techniques to build a communication medium. We will track the progress of RF hardware since 1996 and look at the cutting edge of the pervasive Internet. We will contribute to creating Internet and network access as ubiquitous and analogous as to the way electricity is currently used: users and hardware simply plug in!

You may be creating a network for specific devices or to control equipment in a building. You can have a multi-use network segmented with similar devices on a subnet. This keeps communication traffic limited to only those nodes involved in the exchange. The way you visualize the topology of your network might be a simple line, point A to point B because you just want to connect to the Internet and the nearest broadband connection is way over *there*. Your mental picture might look like a bunch of ever-larger concentric circles with yourself at the center. Your stuff and friends are located in the center rings while the outsourced tech support person in India is very far away at the largest circle. But is he really at the far edge of the network?

If I live very close to D-Link headquarters in Irvine, California and I need technical support, I dial a local phone number and speak with a person in a foreign country. I notice no delay, and the only clue as to her location is her accent. If I click on their Web page's technical support link, I am instantly directed to a server that can be located anywhere in the world. As you begin to create your community LAN, remember to take a deep breath, a step back, and consider that the place you are wiring or unwiring is this planet. You are a member of the technical volunteers tasked with the duty to solve the details and minutia of each node's creation. It doesn't matter if you have a 144 Mbps fiber backhaul over a 20-mile link if I come to your house and I can't print. I'm not going to be impressed if I can't use my mail server. I will not think you are enlightened if my connection prevents me from engaging in communications with your neighbors, or worse, transferring a file to your computer over the local link.

It never ceases to amaze me that when I am working in the same room with other information workers (responsible for the very network we are working on) and need to share a document, I must invariably e-mail the file as an attachment. Often, I must log in to my Web-based junk e-mail account because I can't use the local mail server. This creates needless traffic and problems with large attachments. When I complain, the resident geek will list seven ways to work around the problem. I have these tools as well, but we are trying to build networks for *people*, not machines. I really do want my Mom to enjoy a connected world.

For example, in 1982, I worked on a project for Olivetti. It involved a campus network and we had linked the telephone PBX to the LAN. The LAN knew where you were on campus, and if a phone near where you were standing rang, it was always for you! If someone was trying to call you and you were walking down a hall, the phones in each office would ring as you passed each door. You would pick up any phone and it was a call for you.

What have we done with ubiquitous, pervasive connectivity since 1982?

In this chapter, you will learn about:

- Extending a wireless network using non-802.11 standards

- Terms and protocols used in multihop networks

- Mesh vs. WDS

- Routing protocols

- Community networks

Preparations for the Hacks

This chapter differs from the rest of the book because networks that scale to the point where you need a mesh typically require large amounts of hardware and the experience to deploy it. This limits the average individual or even users group's ability to gain experience with many hardware/software configurations in the mesh world. One box just doesn't cut it. Additionally, the requisite laboratory needed to test such a network requires a large area with at least 50 *willing* users to truly test the efficacy of a solution. The Frequently Asked Questions section of the LocustWorld Web site (www.locustworld.net) states that a mesh network costs between $8,000 and $16,000 to start, and that it would become cost effective at about 50 users. Indeed, SoCalFreeNet.org has only recently reached critical mass, in terms of deployment, where it seems practical to consider an alternative to traditional 802.11-compliant deployments. The reader should not be discouraged, however, as many of the mesh operating systems can be loaded onto the hardware mentioned in other chapters. The Soekris boards populated with radios like Senao and Orinoco are commonly supported. This is true for commercial products such as TurboCell as well as those in the public domain like GNU Zebra. Another networking standard in the public domain published by the National Institute of Standards and

Technology (NIST) is AODV (Ad hoc, On demand, Distance Vector). This popular standard is employed by LocustWorld and is used for long-haul backbones to bring broadband to rural areas.

The Basic Definitions

Wireless Network = Network = Corporate LAN = Community Network

This is a subjective definition of what I believe to be the future of any but the simplest implementation of wireless technology. I think my co-authors would agree that I have a somewhat different concept about how and what we deploy as a group that provides free Internet access. This is a good thing, and one of the reasons why I joined SoCalFreeNet.org. I wanted to expose myself to the whole grassroots adventure of open-source software and homebrew hardware. Long live SoCalFreeNet.org! With that in mind, bear with me if this chapter is a bit on the narrative side and has you chomping at the bit. I know you can't wait to dig in! Be careful of what you wish for, because this chapter will use everything you have learned in previous chapters and then some. Soon, you will be ear deep in hardware!

NEED TO KNOW...WHAT IS A COMMUNITY NETWORK?

A community network is very similar to a corporate LAN with ubiquitous, pervasive, wireless broadband Internet access as just one of the services provided. Today, the effort required to create a full-blown 10/100BaseT network is minimal. Indeed, all the ruckus about loading a specific operating system (OS) with specific radios, with a router that presents a splash page, is to *avoid* the use of traditional networking gear! Many of you are concerned with the cost of implementing the hardware required to host the usual network services. The license fee associated with the use of a true server OS is just one of the concerns with respect to the high cost of network deployment. Consider that many user groups and individuals use a multifunction AP to provide several services such as Dynamic Host Configuration Protocol (DHCP) and Network Address Translation (NAT). The next step is to provide a splash page via port 80 HTTP redirection. Often, a more expensive solution is deployed using a Soekris kit running additional software in addition to the AP. Today, local computer stores sell a sub-$200 computer that can turned into a full-blown Linux server. You can optionally install something like the freeware version of the Public IP Wi-Fi Gateway found on the ZoneCD to complete a similar suite of services (see www.publicip.net for more details). Of course, you can't hoist it up a pole, but once a group has deployed one server, cloning the system is very easy and economical. Providing the necessary space in a wiring closet or cabinet prepares the system for all of the network's future needs. Your T1 line terminates there, and soon it can accommodate that fiber link. Segmentation of services for different groups (open network access versus controlled staff access), scalability, and remote monitoring/control all terminate at this location.

If we are successful, many people will use our networks. My point is that a little commercial thinking will prepare us for the broader mix of different users who will use that system. Second- and third-generation users will be less concerned about what hardware is used, oblivious about the technology incorporated to bridge the last mile, and much more demanding in terms of what applications run on that network.

Many of you sit in front of a computer all day connected to a corporate LAN, which may not be very exciting, but a LAN *can* be fun when you control what services are provided. I suggest that we all install 802.11g APs at the minimum and start using those four ports and the 30 or so wireless clients most APs can support. The best and cheapest part of your AP is the 100 Mbps connection on the local LAN. Find out what common ground exists in your community. Share that large format printer! Get the neighborhood pilot to offer flight lessons on your LAN using MS Flight Simulator. Is water conservation an issue in your part of the country? How about a Web page showing how to landscape your property with indigenous plants? Your community network can create the next generation of responsible NetCitizens, new members of local users groups, and future CTOs and IT managers.

Once this plumbing is installed, you can hook into community networks in other parts of the country. Start your own phone company. Voice over Internet Protocol (VoIP) is ready now. Broadband will get cheaper, and fiber is just around the corner. Check out the real-world examples later in this chapter.

What SoCalFreeNet.Org is really doing is very similar to how the Romans created a great city. The Romans provided its citizens with *free* water via an intricate viaduct and plumbing system. It is what made Rome such a great place to live. Once they had a broadband backbone (a lot of water hauled in over the viaduct), they could provide a "last mile" solution throughout the city by placing pools and fountains on nearly every street corner.

SoCalFreeNet.Org convinces people in a community to allow access to property they own or rent. Those who can afford it (usually a property owner) help to pay for the equipment and the recurring backbone costs. The pitch is the same as it was in ancient Rome. It improves the quality of life in that community. It doesn't hurt property values either. Of course, our volunteers contribute the labor.

Now that the plumbing is in place, if I visit your house and am thirsty, you graciously give me a sip of water. It's free! The same is true of a guest in your neighborhood. Internet access is free! If you visit my house, you may take a sip of my broadband access, for free. I can do the same when I visit your community. It's the civilized thing to do.

It is helpful to know a little bit about network topology if you intend to deploy more than one wireless node. I note the progress in how we deploy LANs to emphasize the changing landscape that always happens with technology. I first worked on a network that had a rigid cable the size of your garden hose. It was so tough to route through a building that one simply bolted it to exterior walls. To add a workstation, you literally "tapped" it with a fixture that punched a hole in cable. I don't think anyone is going to recommend a new deployment of ThickNet today.

Next came ThinNet. The cable was thinner and flexible, but one still added computers in a long serial line. You could run the stuff through walls, but you had to cut and terminate the cable wherever a host joined the network. The first transport protocol that ran on it was 10BaseT. ThinNet is still occasionally used for installations that have a lot of RF noise and where bandwidth demands are modest, as in factory control.

Today, the most common "best practice" approach is the star pattern where people use simple Unshielded Twisted Pair (UTP). UTP is easy to pull through the building and each computer gets a home run to the wiring closet. It is easy to expand because the backbone can be linked from wiring closet to wiring closet. New switches or hubs can be added, and computers are easily plugged into a nearby wall jack.

It is necessary to understand how all of this works so that you can plug in a radio. Even if you plan to use one radio, it must hit the copper at some point. That point is identified as an "edge node" for an "edge device" by your ISP. You are hanging your radio onto their network. That network, in turn, connects other communities. Paranoia, ignorance, and sloth have created an atmosphere of distrust, and most service providers prevent you from seeing any other guest on that network. It doesn't have to be that way. Again, we can learn from the corporate LAN. As with any other important transaction, one simply needs to know: who did what, and when.

Even the most casual connection will address the issue of authentication. You share that broadband connection with your neighbor because you know her. It is more of a problem to authenticate a user who is unseen, but that issue must be addressed. Just as hardware, topology, and software change, we must accommodate the rapidly broadening world and all its rapid changes. Splash pages, disclaimers, and access control are all part of the wireless picture. Cheap or free community wireless carries all the responsibilities found in any public project. That AP had better not fall on my head, the RF signal had better not interfere with my AP (or God forbid, my TV), and don't let a terrorist plot an attack over your connection!

Here then is a definition of *mesh* I found on the TechTarget and Telecom Glossary 2K Web sites:

> "A mesh network is a network that employs one of two connection arrangements, full mesh topology or partial mesh topology. In the full mesh topology, each node is connected directly to each of the others. In the partial mesh topology, nodes are connected to only some, not all, of the other nodes."

Wireless Distribution System

As explained in the *Wireless Networking Starter Kit* by Adam Engst and Glenn Fleishman (www.icsalabs.com and www.wireless-starter-kit.com), in the casual sense of the definition, a Wireless Distribution System (WDS) is a form of mesh. Notably, WDS has been part of the 802.11b specification since 1999, but only recently have consumer Access Points (APs) implemented the option. This makes WDS one of the least expensive ways to try a distributed wireless cloud with a possible solution to non line-of-sight (LOS) connectivity. Protocols used are those familiar Ethernet standards as found in a network switch where the switch routes packets based on MAC addresses assigned to ports. Each AP keeps a list of each associated computer's MAC address as though the AP was a port on a switch, forwarding the packet to the target computer either on the same port or to the computer on a different "port" (AP) even if an intermediate "port" (AP) must act as a middle man to pass the message.

This has two implications. Imagine you are connected to AP 1 and want to send data to AP 3 but do not have an LOS view of AP 3 because of a building or mountain blocking your view. A possible solution is to place a third AP in a location visible (LOS path) to both AP 1 and AP 3. Now your distribution system updates a list of computers connected to each AP and distributes that list to AP 1, 2, and 3. AP 2, the middleman, can see the other two APs and receives a packet from AP 1 with a destination to the computer on AP 3. Acting just like a switch, AP 2 forwards the packet to the unseen AP 3. Only one of the three APs requires a backhaul connection to the Internet. Everything is good so far.

The second implication addresses the issue of bandwidth loss through propagation. How many hops can you make before your net throughput drops below acceptable limits? Consider that 802.11b operating at the maximum rated 11 Mbps has a net throughput of around 3 Mbps; two hops (four APs) can reduce throughput to <1 Mbps. The APs update those MAC lists even if users are not actually accessing the system.

The 802.11 specification for WDS specifies the MAC address of the origination and destination computer with provisions for two additional addresses designed to move the packet to the closest destination port (AP). This is fine if you only require Internet access at or below T1 speeds. It is a poor choice for a wireless network. (See previous definition.) You could use 802.11g to increase throughput, but the ability to scale is obviously limited. As of this writing, Broadcom has the most widely used version of WDS. Nevertheless, most solutions are vendor specific. That is to say that a successful deployment will most likely involve a single vendor's equipment. Fortunately, CPE (Customer Premise Equipment) is not affected by the choice of AP. Scalability is the boasting right of mesh.

I define a true mesh as a network with more than three nodes having a connection schema that can process a communication link through more than two hops. There are many ways to do this, and most protocols begin with a description involving a linked "pair" as the basic (smallest) unit. These "pairs" can be part of a large number of nodes on a network where every node can "see" every other node. More likely, a group of geographically located pairs forms a cloud and only a single AP in the cloud can "see" a distant cloud consisting of a number of different pairs. This is how a mesh network scales.

Indeed, mesh itself is a collection of solutions to many of the problems associated with monolithic wireless networks. You may already be familiar with the "hidden node" problem. This occurs when node A is linked to node C and doesn't realize node B is talking to node C. Node A cannot see node B, but C can see both A and B. This can happen because there is an obstruction between node A and B, or because node B is close to node C but too far away from node A to create a useful link.

You can solve this hidden node problem by writing a table into every node, listing every other node and its location. Now all nodes are aware of each other, but they still are clueless as to whether an unseen pair is talking. The clueless node then talks over the current conversation, which results in lost packets. Now we need a rule that requires a node to ask permission to speak, RTS (Request To Send), and another rule notifying other nodes that it is listening, CTS (Clear To Send).

Creating rules about how and when a node communicates will contribute to system overhead. Choose the wrong protocol and as your network scales, all the available bandwidth is consumed by the system, leaving too little bandwidth to carry actual files and user data. Each vendor addresses different aspects of networking and RF problems. As you run up against a specific problem, the proper solution will not only increase the likelihood of success, but will also save you money.

Real-World Examples

Three residents living in a 50-unit apartment building asked me to provide broadband Internet access. When I approached the owner of the property, he flatly refused to allow me access to the building. Belair Networks (www.belairnetworks.com) markets a solution to lower the cost of network deployment in a multistory building. If you understand the network topologies outlined, it would be prudent to do a cost analysis of the required wiring, switches, and installed wall plates. Wireless is an attractive solution because you can surround the building with an RF cloud. Computers inside the building need only have a client radio within range of the cloud.

My solution was similar. I found locations to mount my APs across the street from the apartment building, aiming my directional antenna at the residents' units. If a computer received a weak signal, I added a directional antenna pointed at my AP across the street.

Belair Networks has made such an installation much easier. Using a mesh technique to get the backhaul around the building where the primary AP cannot see one or more radios surrounding the building, they provide a kit containing everything to make that possible. This includes all the required interfaces for the backhaul, including fiber and Ethernet, mounting hardware for telephone and street light poles, and, most importantly, an autodiscovery setup that configures the network for optimal coverage throughput. Again, the computer inside the highrise need only point an antenna out the nearest side of the building. This is perhaps one of the very best applications for mesh, as it avoids all of the pitfalls of multiple hops, RF interference on adjacent channels, and costly building wiring and buildout. The RF signal is directionally constrained and attenuated by the very building serviced! Simply put, this is a great way to cover a building with RF. I add that Belair Networks does offer other solutions that you can review at their Web site.

Example Two: LocustWorld Mesh Networks

LocustWorld (www.locustworld.com) makes one of the easiest to use hardware and software products for mesh networks. Their Web site claims that this is the most popular form of mesh in the world, especially in Europe. For those of you unfamiliar with Linux and the methods to load an OS image onto a computer, LocustWorld sells the LocustWorld MeshBox. It is a ready-to-use AP loaded with their modified version of the 2.4.19 Linux kernel as the POSIX OS. The MeshAP routing software runs as an application with a GUI and a CLI. The box contains a single-board computer with no moving parts. The MeshAP routing software supports 802.11b with a throughput of about 5 Mbps. When a regular 802.11b AP runs in infrastructure mode as a base station associated with a client that is close enough to connect at the full 11 Mbps connection speed, the throughput after overhead is about 3 Mbps. The difference lies in the operational mode of the MeshBox. It runs in ad-hoc mode similar to a peer-to-peer network. Beacons are turned off and other 802.11b-compliant features are disabled. Each AP does double duty as a router and repeater, thereby extending the range of the wireless network. This reduces operational overhead, leaving more bandwidth for users. Rumor has it that a 5 GHz version will soon be supported.

Each AP self organizes its routing table so that it can pass packets to the nearest neighbor. Removing or relocating an AP updates the routing tables, deleting old routes as needed. Routing tables are not passed from one AP to another. This means it is possible that an AP does not know about the presence of its neighbor's neighbor. This method does not fully address the "hidden node" problem. A careful survey and some testing are required before deploying this solution in a topology where not all radios can see their counterpart's pair in a given RF cloud. When this happens, the mesh network might split or segment. Each part is then unaware of the other and communication between segments ceases.

Good examples for the appropriate use of this technology would be for a triage situation such as smoke jumpers deployed from a helicopter to fight a fire with a clear LOS or small public events like a street festival.

Another good point is that the MeshAP routing software is available as an open-source project based on open standards supporting a number of different radio cards so you can brew your own LocustWorld mesh solution.

WARNING... CHOOSE YOUR HARDWARE PLATFORM CAREFULLY

If you plan to build your own AP, make sure the hardware is up to the task. If the AP is located in a difficult-to-reach location, a fan that fails will require considerable effort to replace. Consider the weather patterns of your location. While it seems obvious to plan for extremes in temperature, you might forget to check other operating specifications such as altitude and humidity. Many electronic devices are not rated above 8,000 to 10,000 feet, as I found out in Colorado. Operating an AP above 12,000 feet resulted in erratic behavior that was difficult to diagnose. Replacing an AP at an altitude beyond the ski lift in Aspen requires a very serious hike.

Summary

In this chapter, we reviewed the concepts and applications of mesh networking. As opposed to standard 802.11 networks, which typically use an "infrastructure node" with APs and clients, mesh networking operates with a completely different model. No APs are used. Each node is "autonomous" and communicates with its peer nodes. Special routing protocols help move packets around the network.

Community wireless networks have deployed mesh-type networks using a special distribution called LocustWorld from http://locustworld.com. A number of commercial products also exist from companies such as Belair Networks, Tropos, and Firetide.

Mesh networks have extraordinary capabilities and should always be considered a potential solution for large-scale metropolitan area networks (MANs). This chapter is intended as a primer to help you understand the basic concepts so that you can explore this exciting area of networking further. The following URLs are a good starting point for your journey.

Additional Resources on the Web

- **Mesh Networking Project in Germany** http://free2air.org/
- **MIT Roofnet** www.bawia.org
- **www.pdos.lcs.mit.edu/papers/grid:bac-meng.pdf**
- **LocustWorld** www.locustworld.net
- **AODV** http://moment.cs.ucsb.edu/aodv-ietf
- **Mobile Mesh** www.mitre.org/work/tech_transfer/mobilemesh
- **TBRPF** www.erg.sri.com/projects/tbrpf

Companies developing mesh products:

- **www.meshdynamics.com**
- **www.strixsystems.com**
- **www.tropos.com**
- **www.4g-systems.biz**
- **www.firetide.com**

## Part IV

Antennas and Outdoor Enclosure Projects

## Chapter 10

Antennas

Topics in this Chapter:

- **Before You Start: Basic Concepts and Definitions**
- **Building a Waveguide "Coffee Can" Antenna**
- **The Future of Antennas**

Introduction

Whether your wireless system is a simple home office setup or a large-scale outdoor wireless network, the antenna system is the most important, but often most overlooked, aspect of the deployment. The antenna system is often the "make or break" factor for a successful wireless transmission. Poor antenna selection or design can lead to frustration and intermittent connectivity problems. This translates into poor throughput performance and frustrated wireless users.

In this chapter, we explore the important issues surrounding antenna selection for any 2.4 or 5 GHz unlicensed wireless system. You'll get all the information you need to achieve the best performance possible. We'll examine many types of commercially available antennas, and you'll see how you can build your own antenna as an alternative, using inexpensive materials from the local hardware store.

At the conclusion of this chapter, you should:

- Understand the different decibel (dB) measurements and how to use them correctly
- Understand the Federal Communications Commission (FCC) rules and power output requirements for operation in the 2.4 and 5 GHz bands
- Understand the different types of antennas and use Figures of Merit (FOMs) to determine the best choice
- Understand the "3 dB" rule and its importance in determining system performance
- Understand "wavelengths" and use a formula to determine how long an antenna needs to be for a given frequency
- Determine system performance requirements using the link budget calculation
- Estimate the "fresnel zone" to determine proper antenna height and orientation
- Understand how different building materials can affect performance
- Be aware of safety precautions such as lightning arrestors, proper grounding techniques, and building code adherence

Before You Start:
Basic Concepts and Definitions

Before you can install and/or construct any antenna, there are several terms and calculations with which you should be familiar. While a degree in physics is not necessary, a basic understanding of physics is helpful.

An antenna is simply a passive transducer that radiates energy (*gain*) into space. Antennas do not actually amplify the signal; they simply change the shape of the energy pattern being radiated. You should be able to select or construct a basic antenna for your use once you understand the basics of antenna design, construction, and operation.

The decibel is the most important unit of measurement when looking at antenna performance. The decibel (or dB) is the basic unit used for radio frequency (RF) power measurement. Table 10.1 lists decibel power levels in relation to wattage levels.

Table 10.1 Transmit Power in Decibels

Watts	Decibels
1/1000	0 dB
1/100	10 dB
1/10	20 dB
1/4	24 dB
1/2	27 dB
1	30 dB
2	33 dB
5	37 dB

We use decibel measurements because signal strengths vary logarithmically, not linearly. A logarithmic scale allows simple numbers to represent large variations in signal levels. You'll see it's also very useful in calculating system gains and losses. In the following sections, we've included brief definitions of all the terms we'll be using in this chapter:

- **dB** Decibel. The basic unit of measurement that represents the ratio of two signal levels.

- **dBm/dBW** Decibel milliWatt. This measurement is used to represent power, with 0 dBm defined as 1 milliWatt. For larger signals, there is also dBW, a reference to 1W. Small signals are represented as negative numbers, (for example, −95 dBm). When referencing commercial Wi-Fi devices, power output is normally given in dBm. Many WLAN PCMCIA cards and some Access Points (APs) have a power output of +17dBm (50mW). There is also usually a Receive Signal Sensitivity Indicator (RSSI) measurement listed in dBm (for example, −95dBm).

- **dBd** Decibel dipole. The output power (gain) an antenna has over a dipole antenna at the same frequency. A dipole (two-pole) antenna is a 1/2 wave antenna used as a reference against all other antennas. It is a reference known as 0 dBd (zero decibel referenced to dipole). The dBd measurement is usually used only with antennas below 1 GHz.

- **dBi** Decibel isotropic. This measurement is used for antennas above 1 GHz. A dipole antenna has 2.14 dB higher gain than the 0 dBi dipole reference. If antenna gain is given in dBd, not dBi, add 2.14 to convert to the dBi rating.

NEED TO KNOW...RF POWER

There are several basic rules that you should know when working with antennas, RF power, and expected signal strength. The "3 dB" rule is perhaps the most important rule when dealing with RF (signal) power. It states that for every 3 dB increase in level, the power is *doubled*. For every 3 dB decrease, the power is cut in half. Similarly, every 10 dB increase in level is 10 times the power, and every 10 dB decrease in level results in 1/10 the power. This is sometimes referred to as the "rule of 3s and 10s."

Once you understand the different decibel measurements, it is easy to understand Figures of Merit (FoMs) when working with antennas. FoMs are attributes that describe an antenna's performance characteristics. The FoMs are listed as part of every antenna's specifications. Important FoM attributes like *gain* and front-to-back ratio are listed in dB or dBm. There are many other RF terms and figures that use decibel references and values (these terms are explained in greater detail later in this chapter). Once you are familiar with FoMs in general, it will be easy to recognize the important features of antennas and choose the best antenna for your application.

Effective Isotropic Radiated Power (EIRP) is defined as the power found in the *main* lobe of the antenna relative to an Isotropic radiator with 0 dB of gain. The EIRP is calculated by taking the antenna gain (in dBi) plus the power (in dBm) inbound from the transmitter. For example, a 9 dBi antenna fed with 26 dBm of power would have an EIRP of 35 dBm.

9 dBi + 26 dBm = 35 dBm (3.2W)

The chart in Figure 10.1, known as a *Smith* chart, shows the propagation area of a Yagi antenna. A Smith chart is included with any antenna specification and represents the radiation pattern of the antenna. It also shows the front-to-back ratio, and the "side lobes," which are the smaller, less powerful radiation patterns on each side of the main lobe.

Figure 10.1 Representation of a Unidirectional Yagi Antenna Radiation Pattern

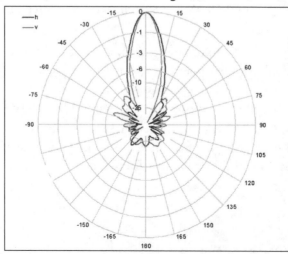

Courtesy of Pacific Wireless (www.pacificwireless.com)

The top pattern represents the main lobe and transmit gain. The lower pattern the back lobe. The difference (in dB) between the front and back lobe is called the front-to-back ratio.

NOTE... A WORD ABOUT ANTENNA GAIN AND COVERAGE

Since the EIRP is in the *main* antenna lobe only, antenna selection is critical.

When using a high-gain omni antenna (8–12 dBi), the propagation angle is very flat and narrow. Placing the antenna too high will cause the main lobe to pass over the intended target antenna. The irony here is that height is required to clear obstructions, a.k.a. Line-of-Sight, from the Wireless Point of Presence (WiPoP) path to the receivers. Higher gain omni antennas have a flatter, "pancake" shape, while lower gain omni antennas tend to have a wider "donut" shaped pattern.

It may be necessary to use a unidirectional antenna and "down tilt" that concentrates the energy (signal) in a more focused area. Unidirectional antennas direct energy in one direction by radiating the entire signal in a concentrated area instead of 360 degrees like an omni. Table 10.2 lists antenna types and associated values in dBi (gain). Figures 10.2 through 10.6 are images of these antenna types.

Table 10.2 Typical Antenna Types and Gain Values for Off-the-Shelf Antennas

Antenna Type	Gain (dBi as we're dealing with >1GHz) Freq.
Unity gain Omni	0 dBi
Low Gain Omni	2–6 dBi
High Gain Omni	8–12 dBi
4 x 6" Panel (Unidirectional)	7 dBi
Small Yagi	10 dBi
8"– 10" Panel (Uni)	13 dBi
12" Panel (Uni)	16 dBi
Long Yagi	16 dBi
18" Parabolic Dish	19 dBi
18" Diagonal Mesh/Grid Antenna	21 dBi
24" Diagonal Mesh/Grid Antenna	24 dBi

Figure 10.2 8 dBi Omni

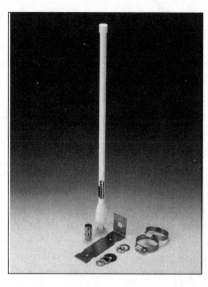

Figure 10.3 8 dBi Uni (Panel)

Figure 10.4 Large Omni

Figure 10.5 24" x 36" Mesh Grid Antenna (21 dBi)

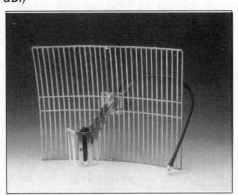

Figure 10.6 Yagi (12 dBi)

NOTE... INTERESTING ANTENNA

An interesting antenna type has been developed by cantenna.com. This "super cantenna" resembles a Pringles can antenna, is linearly polarized, and features a gain of 12 dBi and a beam width of 30 degrees. You can learn more about this innovative, low-cost product at www.cantenna.com.

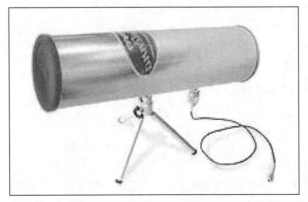

Federal Communications Commission

A common misconception when using 'unlicensed' equipment is that there are no rules covering the operation of such gear. While there are no license requirements, the FCC does have some regulations with respect to the maximum power output levels when using unlicensed equipment. Part 15 of the FCC's rules for radio equipment lists the specific power requirements. We discuss the pertinent limitations in this section.

The FCC has relaxed the rules on EIRP limits for Point-to-Point (PtP) systems. This has increased the choices of antennas and extended the range of PtP systems. The EIRP for a 2.4–2.5 GHz PtP system is now 36dBm (an amazing 4 watts!) We must calculate a link budget to determine the total EIRP, and remain in FCC compliance. The FCC allows *only* 30 dBm (1W) EIRP for Point-to-Multipoint (PtMP) communications. This limits the antenna choices and makes the calculation of system output very important. However, for most off-the-shelf commercial equipment using attached antennas, the output is 50–200 mW. This coupled with a 6 dBi antenna is well below FCC limits. Using the previous charts and remembering the rules will help you calculate power levels and remain in compliance. A good rule to remember for 2.4 GHz PtP systems is that (at maximum power output levels) for every 3 dBi of antenna gain over 6 dBi, the transmitter power output must be reduced by 1 dB. For 2.4 GHz PtMP (at maximum power output levels), every 3 dBi of antenna gain over 6 dBi must be met with a 3 dB reduction in transmitter power.

The 5 GHz band has various output power limits. The limits depend upon the sub-band within the 5 GHz band in which you're operating. The lower portions of the 5 GHz unlicensed band are between 5.15 and 5.25 GHz The output for these devices is fixed at a maximum of 50 mW. The 5.25–5.35 GHz middle sub-band has a power limit of 250 mW.

The 5.725–5.825 GHz upper band is normally used for high bandwidth (T-1 , OC-3) transmissions associated with microwave radio. This band has most recently been adopted by many Wireless Internet Service Providers (WISPs) as a high data rate "backhaul" solution. This removes congestion from the 2.4 GHz (DSSS) frequency band and allows much more bandwidth (and more users) to be concentrated for transmission.

The Link Budget is the calculation of the losses and gains (in dB) for the complete RF system, and is determined using a simple formula that combines all the power and gain figures for both sides of a link.

Link Budget = P(t) + TX(G) + Rx(G) + Rx - Path Loss

Where:

- P(t) = power of transmitter (e.g., 17 dBm)
- TX(G) = transmit antenna gain (e.g., 6 dBi)
- RX(G) = receive antenna gain (e.g., 6 dBi)
- Rx = Receive Sensitivity of receiver

The numbers are the *gain* figures used in a link budget. We will also look at the loss or attenuation levels—caused by cables, connectors, and so forth—that *must* also be factored into the final Link

Budget calculation. (A good online calculator can be found at www.afar.net/RF_calc.htm and www.qsl.net/pa0hoo/helix_wifi/linkbudgetcalc/wlan_budgetcalc.html)

Path loss, the amount of loss in dB that occurs when a radio signal travels through free space (air), is also known as Free Space Loss (FSL). FSL can be calculated using the following formula:

FSL (isotropic) = 20Log10 (Freq in MHz) + 20Log10 (Distance in Miles) + 36.6

Additional factors you should consider when determining your link's requirements:

- **Radiation pattern/propagation angle** The propagation angle is given in degrees and denotes how much area in degrees an antenna broadcasts its signal. Example: Vertical angle = 45 degrees, Horizontal angle = 7 degrees. Search the Internet for various antenna manufacturers to find examples of Smith charts that represent various propagation angles.

- **Polarity** All antennas have a "pole" (short for polarity), which can be horizontal, vertical, or circularly polarized. Polarity indicates the angle of the RF wave's propagation in reference to an H/V/C plane. You *must* insure that all Wi-Fi systems you want to communicate with have antennas on the same pole. The difference in H/V poles (if for example, one antenna is horizontally polarized and the other is vertically polarized) is a loss of 30 dB.

- **Vertical/horizontal beam width** This is the angle of the RF "beam" referenced to the horizontal or vertical plane. Typically, the higher the gain, the more focused (narrow) the beam. Example: A 24 dBi antenna commonly has an 18-degree beam width, vs. a 9 dBi antenna, which will have a 45- to 60-degree beam width.

- **Fresnel zone** The Fresnel zone is the propagation path that the signal will take through the air. The Fresnel zone can be determined using the formula below. The Fresnel zone is important when installing Line-of-Site equipment, because if the Fresnel zone or any part of it is obstructed, it will have a direct and negative effect on the system connectivity.

Fresnel Zone Calculation = 72.1 * SqrRoot(Dst1Mi * Dist2Mi / Freq (in GHz) * Distance-in-Miles

You can find a good online Fresnel zone calculator at www.radiolan.com/fresnel.html.

- **Front-to-back ratio** An antenna's front-to-back ratio is typically given in dB and denotes how much signal is projected behind the antenna, relative to the signal projected in front of the antenna (in the main lobes). The lower the front-to-back ratio, measured in dB, the better. The reason is that you don't want excessive signal propagating from the rear of the antenna.

- **Link Margin** The Link Margin, sometimes called System Operating Margin (SOM), is the minimum difference between the received signal (in dBm) and the sensitivity of the receiver required for error-free operation. In many systems, this is also referred to as the Signal-to-Noise-Ratio (SNR).

Table 10.3 lists Fade Margins for various link distances.

Table 10.3 Fade Margins for Various Link Distances

Distance (Miles)	Conservative Fade Margin (dB)
0.5	4.2
1	7.5
2	10.8
3	12.75
4	14.1
5	15.2
10	18.5
15	20.4

In many newer radios, a Signal to Noise Ratio (SNR) specification is used instead of the RSSI reading/measurement. Motorola's 5 GHz Canopy system requires only 3 dB SNR to achieve connectivity, while Alvarion's EasyBridge 5.8 GHz system expects a minimum 10 dB SNR for connectivity. Several good Web sites provide calculators for Fresnel Zone, Fade Margin, and Path Loss:

- www.zytrax.com/tech/wireless/calc.htm

- www.dataradio.com/mso/tsan002rf.xls

- www.andrew.com/products/antennas/bsa/default.aspx?Calculators/qfreespace.htm

NEED TO KNOW...THE BIGGER THEY ARE, THE FARTHER THEY CALL

Size does matter! It may be necessary to increase the size of your antenna if you find that you can't quite get the desired distance or throughput from your link. Remember the "6 dB" rule when thinking about antennas (size), propagation distance, and path loss. The rule states that each time you double the distance from transmitter to receiver, the signal level decreases by 6 dB.

Attenuation in Cables, Connectors, and Materials

Attenuation is the reduction in signal due to cable length, connectors, adapters, environment, or building materials. Often, indoor wireless systems will suffer extreme attenuation due to metal cross members or rebar within walls. It is important to consider the type of building materials used for either indoor systems or systems where client antennas are mounted indoors while AP antennas are outdoors at a distance. It is also important to take the figures for cable and connector loss into account when calculating your link budget.

Table 10.4 lists common building materials and the expected loss in dB.

Table 10.4 Attenuation Factors for Various Materials

Material	Attenuation Factor/dB Loss
Plasterboard wall	3 dB
Glass wall with metal frame	6 dB
Cinder block wall	4 dB
Office window	3 dB
Metal door	6 dB
Metal door in brick wall	12.4 dB

The most common cables used in unlicensed wireless include:

- **RG-58** Commonly used for pigtails and is not recommended for long runs. Loss at 2.4 GHz per 100 feet = 24.8 dB.

- **LMR 195** Identical in gauge to RG 58, but with less loss. Loss at 2.4 GHz per 100 feet = 18.6 dB.

- **LMR 400** Used most commonly for antenna runs over 6 feet. Loss at 2.4 GHz per 100 feet = 6.6 dB.

- **LMR 600** The best, but also the most expensive cable. Loss at 2.4 GHz per 100 feet = 4.3 dB.

The loss quoted for any cable specification is generally per 100 feet. The loss factor is important to remember when installing outdoor systems. For both cables and connectors, the loss factor is commonly listed as "insertion loss." A good online cable loss calculator can be found at www.timesmicrowave.com/cgi-bin/calculate.pl.

Figures 10.7 through 10.11 are examples of connector types used in unlicensed wireless systems. In most cases, it is assumed that the loss per connector is between .2 and 1.0 dB. Many people use .5 dB of loss per connector as a general rule of thumb. If a connector is suspect and produces more loss, it is either of poor design or is faulty.

Figure 10.7 "N" Type

Figure 10.8 SMA

Figure 10.9 MMCX

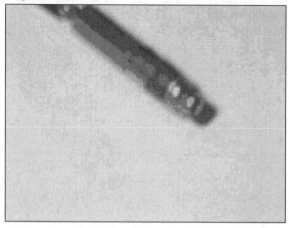

Figure 10.10 TNC

Figure 10.11 Reverse Polarity (R/P) TNC

System Grounding and Lightning Protection

Since an antenna is a metal object with a corresponding wire connection and is elevated several feet in the air, it unfortunately makes an excellent lightning rod. It is always recommended that you use both an earth ground and a lightning arrestor when installing antennas outdoors. The earth ground should be connected to the antenna mast and the antenna tower to ground electrical charges (lightning). It is also recommended to use a lightning arrestor to protect radio equipment. The insertion loss of a good lightning arrestor is commonly a maximum of 1.5 dB.

Figure 10.12 shows a typical lightning arrestor.

Figure 10.12 Common Lightning Arrestor for 2.4 GHz

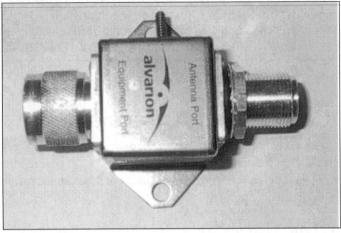

WARNING: HARDWARE HARM

The labeling on the lightning arrestor denotes the antenna port connection and the equipment (radio) port connection. Connecting the device in reverse may result in damage to equipment and systems. It is also quite probable that the system will not work or performance will be severely degraded.

The lightning arrestor should be located between the radio equipment and the antenna. Figure 10.13 is an example of a small unidirectional antenna with jumper cable plus a lightning arrestor and pigtail assembly. This could be mounted on a pole, on the side of an eave, or in conjunction with an outdoor box containing the radio.

Figure 10.13 Lightning Arrestor Mounting Scenario

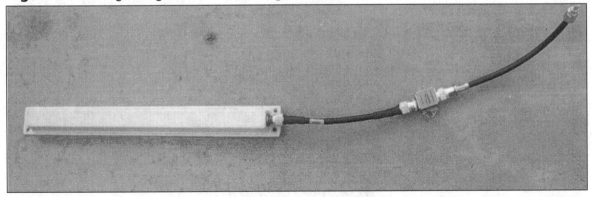

WARNING: HARDWARE HARM

It is always recommended that proper grounding techniques and lightning protection devices be used when installing any antenna system outdoors. Always use caution when installing antennas, especially when using extended masts or building tower sections.

Building a Coffee Can Antenna

If you'd rather not purchase antennas from one of the many commercial options, there are many Do-It-Yourself designs available. For those of you who are interested in experimenting, we'll start with building a coffee can antenna. The coffee can antenna hack we'll be describing here will provide up to 11 dBi of gain at 2.4 GHz.

Preparing for the Hack

Before constructing any antenna, there are two important formulas you need to know. The first is a Frequency/Wavelength formula. For our purposes, we'll use Megahertz instead of Gigahertz.

This tells us the wavelength for our coffee can antenna. For example, if we use 2.45 GHz (the middle of 2.4 GHz band), we get a wavelength of = .4016 feet (984/2450).

The materials required for this hack are:

- Garden-variety coffee can as shown in Figure 10.14 (Folgers or Maxwell House will do). The best cans will be 3 to 3.5 inches in diameter, as long as possible.

- 1.2" brass rod or 12-gauge solid core electrical wire

- Type "N" bulkhead connector

- Four very small nuts and bolts (long enough to extend through the connector and can)

Figure 10.14 Coffee Can

Performing the Hack

To perform the hack:

1. Drill a 1/2" hole, for the type "N" connector.
   If your can has a 3" diameter, the hole should be 3.75" from the bottom of the can.
   If your can has a 3.25" diameter, the hole should be 2.5" from the bottom of the can.
   If your can has a 3.5" diameter, the hole should be 2.07" from the bottom of the can.
   If your can has a 3.75" diameter, the hole should be 1.85" from the bottom of the can.
   If your can has a 4" diameter, the hole should be 1.72" from the bottom of the can.

2. Tin the bulkhead connector by applying a light coat of solder to the "inside" center pin (the opposite side of where the cable is connected).

3. Cut a brass rod 1.2" in length and solder the connector to the brass rod. You can also use solid core 12-gauge electrical wire. Figure 10.15 shows "helping hands," which can be useful when you need an extra set of hands for soldering. Figure 10.16 shows a completed element.

4. Insert the bulkhead connector into the can (the wire/rod portion goes in the can; the other side, where the cable attaches, goes outside the can). Use the four bolts/nuts to secure the connector in place. You may need to drill some small pilot holes in the can to get the bolts through. Figure 10.17 shows a completed coffee can antenna.

Figure 10.15 "Helping Hands" Helpful when Soldering Wire and Connectors

Figure 10.16 The Completed "Waveguide" Element

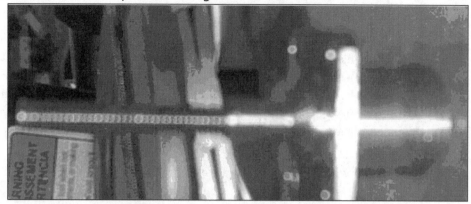

Figure 10.17 The Completed Coffee Can Antenna

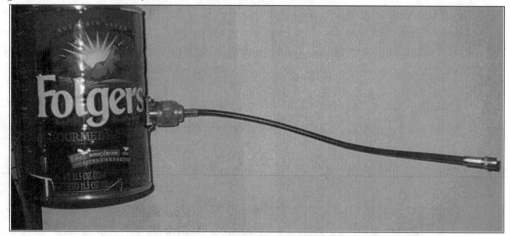

The coffee-can side of the pigtail is an "N" connector, while the other side (for connecting to the radio) is an SMA connector. Various types of connectors may be used depending on the connector interface required by the PC card or subscriber unit.

Need to Know…Save the Jumper/Pull the Pigtail

It is important to remember that most wireless APs will require a short cable commonly referred to as a "pigtail" to interface between the antenna and the AP. This cable is usually 3"–6" in length with connectors on each end. There are several types of connectors used on commercial APs and client cards. It is also sometimes necessary to use a short "jumper" cable between the lightning arrestor or outdoor enclosure and the antenna. These cables should be 6" to 10". Figure 10.18 shows a common 10" pigtail with "N" Connector and MMCX (PCMCIA) Connector. Figure 10.19 shows a 6" N-to-N jumper used between the antenna and lightning arrestor.

Figure 10.18 A Common 10" Pigtail with "N" Connector and MMCX (PCMCIA) Connector

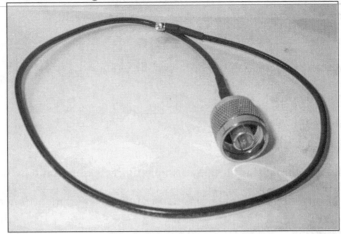

Figure 10.19 A 6" N-to-N Jumper Used between the Antenna and Lightning Arrestor

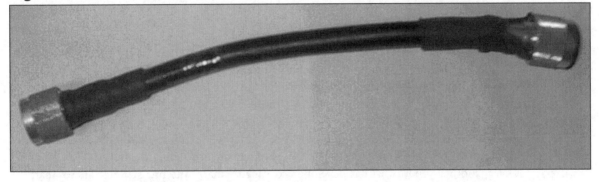

Under the Hood: How the Hack Works

Lightning arrestors are basically voltage "redirectors" that really do not eliminate all electrical charges. However, the standard 1/4 wave stub lightning arrestors from PolyPhaser are the best type for unlicensed wireless in the 2.4 and 5 GHz frequency range. It is important to remember that lightning arrestors are rated for frequency. Always check the specifications for lightning arrestors prior to purchase and installation.

Troubleshooting Common Antenna Issues

It is often necessary to troubleshoot systems when performance falls short of expectations. The following tips will help you determine what the problem(s) might be with poor signal quality, poor throughput performance, or a combination thereof.

When there is no reception, and power and system connections appear correct, some possible problems could be:

- Antenna polarity is reverse of distant antenna

- Lightning arrestor is installed backward

- RF cable has incorrect termination or excessive loss

Poor signal strength on wireless monitor or radio LED indicators could be because:

- Connectors not tight

- Cables poorly terminated

- Lightning arrestor backward

Intermittent signal fluctuations during transmission and reception could be the result of:

- Interferences from friendly or phantom transmitters or equipment (microwave, cordless phone, other APs)

- Multiple antennas on the same polarity—try switching one or alternating antennas to the cross (reverse) pole

The Future of Antennas

Recently, there have been some exciting developments in the field of antenna technology, specifically related to Wi-Fi and the coming WiMax systems. Airgo Networks Inc. (www.airgonet.com) has developed antenna technology based on the yet-to-be-ratified 802.11n MIMO standard. The MIMO acronym stands for Multiple Input/Multiple Output, and uses multiple antennas to increase the range of 802.11 wireless systems. It is designed to increase speed, improve reliability, and reduce interference. These systems (claim to) provide four times (4X!) the coverage area of standard antennas.

Array COM is another vendor that has developed so-called "smart" antenna systems. These smart antenna systems are capable of remote tuning and/or automatic gain and beam width adjustment based on sampled conditions. The following is a list of these antenna types and a brief description of each:

- **Dual polarity antennas** Antennas that are capable of either horizontal or vertical polarity. The antennas typically have separate connectors for both H and V polarity It is not possible to operate at both polarities simultaneously.

- **Multi-gain and variable beam, tunable antennas** A multiple gain, variable beam antenna is capable of operating at various gains, given a desired beam width. Typically, the higher the gain, the more focused the beam width. A common antenna of this type is a TelTek 2304–3. The antenna has settings for 60, 90, 120, beam width. The gain figures rise as the beam width decreases. Example: 24 dBi gain @ 60 degrees, 12 dBi @ 90 degrees.

- **"Smart" antennas** Antennas that adjust automatically to the performance characteristics of the system.

Summary

In this chapter, we reviewed RF Math (rule of 10s and 3s), antenna types, FCC regulations, polarization, Fresnel zones, connector types, and safety issues (grounding and lightning arrestors). We also took you through the steps to build your own coffee can antenna.

Selecting the right antenna for your project is one of the most important steps of any wireless deployment. Antennas do not actually increase the system power. Rather, they merely "reshape" the RF pattern and focus the energy in a particular direction. Antennas are rated with various "gains," as measured in decibels (dB). Use good cables and connectors to help defend against unnecessary signal loss. Thicker, more expensive cables often have the lowest amount of loss.

Always be sure to pay special attention to safety issues. As outdoor mounted antennas are at risk of lightning strikes, make sure to use a lightning arrestor and proper grounding for both your antenna and mast. Be sure to use safety cables for your antennas and antenna masts to make sure that nobody is injured below if a mast were to accidentally come loose or fall.

## Chapter 11

Building Outdoor Enclosures and Antenna Masts

Topics in this Chapter:

- Building Outdoor Enclosures
- Building Antenna Masts

Introduction

The design and implementation of outdoor wireless networks is a challenge that spans many disciplines, from programming, to radio frequency (RF) engineering, to carpentry and metal work. As with any multifaceted project, the final product is only as strong as its weakest link, making even the most mundane detail just as critical as the next. With thoughtful planning and careful implementation of every detail, from the choice of enclosure to the type of fastener used to secure your antenna, you can take steps to help ensure your outdoor wireless networks' dependability regardless of weather conditions.

In this chapter you will learn how to:

- Choose the appropriate enclosure for your needs
- Select proper hardware
- Secure the communications equipment inside the enclosure
- Secure the enclosure itself
- Construct a sturdy mast for your antenna
- Protect your equipment from lightning strikes

Building Outdoor Enclosures

In my years of building community wireless networks in San Diego, my fellow wireless enthusiasts and I have been through many different iterations of outdoor wireless equipment enclosures. These different enclosures span from the infamous "Tupperware Enclosure" to a $70.00 turnkey box made to accept the wireless gear of your choice. Between these two extremes exists a category of enclosures, which meets the needs of the application yet still require some extra attention to bring all of the pieces together.

NEED TO KNOW...TURNKEY ENCLOSURES

"Turnkey" enclosures are made to house specific equipment, to be secured in a specific way, and to be ready to use right out of the box. While these types of enclosures are usually high in quality and easy to install, they are expensive and restrictive regarding what types of equipment can be installed in them and where and in what manner they can be installed. For these reasons, many of the enclosures used in the *SoCalFreeNet* networks are of the variety discussed in the following pages.

Preparing for the Hack

Building an outdoor enclosure can be a very simple procedure involving plastic food containers with holes poked in them, or it can involve elaborate fabrication techniques requiring thousands of dollars worth of equipment. In the following pages, we attempt to strike a balance between these two extremes, constructing a durable enclosure that meets all of your needs while keeping the necessary time and equipment investment to a minimum. The following are some of the basic tools required to perform this hack:

- A basic tool set (wrenches, pliers, screwdrivers)
- A drill with an assortment of fractional-sized drill bits
- A tape measure
- A hacksaw
- Weatherproof silicone sealant

Additional tools, while not required, are extremely useful for this and future projects you may undertake. They include:

- A dremel
- A center punch kit
- A tap-and-die set
- An electric circular saw

Selecting a Raw Enclosure

This section focuses on modifying an existing "off-the-shelf" enclosure rather than building one from raw materials. Most of the enclosures discussed here are described with a National Electronics Manufacturers Association (NEMA) rating. See Table 11.1 for a guide to the NEMA rating system.

Table 11.1 Guide to NEMA Ratings for Outdoor Enclosures

NEMA Rating	Description
NEMA 3	Provides some protection against windblown dust, rain, and sleet.
NEMA 3R	Provides protection from falling rain, but with less overall protection than NEMA 3. This is the minimum NEMA protection value recommended for outdoor use. It provides protection from falling rain, but not necessarily from wind driven rain.
NEMA 4	Provides all of the protection of NEMA 3 plus protection from splashing water and hose-directed water.

Continued

Table 11.1 Guide to NEMA Ratings for Outdoor Enclosures

NEMA Rating	Description
NEMA 4X	Provides the same protection as NEMA 4 plus added corrosion protection.
NEMA 6	Provides protection from the entry of water when temporarily submerged under water at a limited depth.

Selecting a NEMA-Rated Enclosure

Selecting the proper NEMA rating is an important undertaking when planning for your outdoor access point. A selection of NEMA 3 will provide ample protection for most applications. In fact, many of the access points used by SoCalFreeNet are housed in NEMA 3 enclosures. Some have been operational for over a year without ever being re-opened. Of course, this selection makes sense in San Diego, where it hardly ever rains, and when it does, it's usually just falling rain, not the kind of driving rain that can find its way into almost anything. If you live in a region that is prone to heavy rain and storms, perhaps a selection of NEMA 4 or NEMA 4X would be more appropriate for your needs.

NEMA 4 enclosures tend to be plastic and almost completely sealed. If it is determined that this level of protection is required for your project, extra care must be taken to ensure that the methods used to mount the enclosure to your structure and the entry points for the antenna wire and Category 5 cable (Cat-5) (if used) do not compromise the enclosure's ability to keep out water and dust. These types of enclosures are often designed with these sorts of issues in mind, and offer external mounting points and special watertight grommets, which slide over any cables entering the enclosure. If these components are not included, and drilling is required to provide these features, silicon sealant is a good way to restore the enclosure's water and dust resistance. Once the appropriate-sized hole is drilled into the case and the cable is routed through and in place, a liberal application of silicone to both the inside and outside around the cable entry point can restore its protective qualities.

Sizing the Enclosure

The first thing you must decide on is how big you need your enclosure to be. In order to answer this question, you must determine the dimensions of the equipment the enclosure will house. This is a straightforward procedure if you already have the communications equipment in your possession. The simplest method of determining the size requirements for the enclosure is to take your equipment to Home Depot or an electrical supply store to physically determine whether or not it fits.

If you do not have the communications equipment in your possession yet, you can visit the product manufacturer's Web site. Almost all manufacturers have a specifications page, usually in Adobe Acrobat (.pdf) format. There you will find information regarding the physical dimensions of the equipment, the weight, the minimum and maximum temperature thresholds, the power consumption, and other useful information. The dimensions are usually specified in "length by width by height" format. (Information for Soekris and other single board computer hardware can be found in Chapter 4.)

When selecting an enclosure, it is a good idea to add at least one inch to all three dimensions for clearance. This is especially true for the steel NEMA 3 and NEMA 3R electrical enclosures typically found at stores like Home Depot. This is due to the inner flange the lid bolts are secured to (see Figure 11.1). Even though the equipment may match the advertised enclosure's dimensions, the flange may prevent installation.

Figure 11.1 Inner Flange of Steel Electrical Enclosure

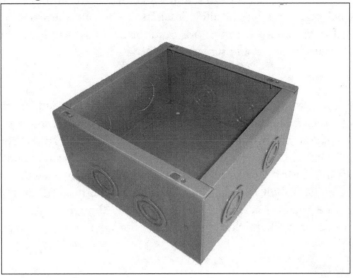

Another sizing issue to keep in mind is the RJ-45 (Ethernet) connector jacks and serial ports usually located on the perimeter of communications equipment. Not only will the wires connected to these ports take up more space than the advertised dimensions of the equipment, but additional space must also be allotted in order to disconnect and reconnect the wires as needed. Oftentimes, once a device is operational, upgrades or diagnostics will require plugging in a laptop or other device directly to the equipment via the RJ-45 or serial port. Members of SoCalFreeNet learned a time-consuming lesson the hard way when these diagnostics required removing the communications equipment from the outdoor enclosure to plug into a serial cable, and then reinstalling it (often in precarious rooftop situations). A little planning can go a long way in preventing headaches down the road.

Another sizing issue to take into consideration is the location of the mounting points necessary for securing the enclosure to the structure. For instance, many outdoor enclosures used in SoCalFreeNet projects are secured to poles using U-bolts. For some applications this became a problem because the ends of the U-bolts protruding into the enclosure were left with little or no room due to the communications equipment covering nearly the entire mounting surface that they both shared. This can be avoided by either selecting an enclosure large enough to accommodate both U-bolts and equipment, or by using one of the side walls of the enclosure to U-bolt the enclosure to the mast. While the results are not as aesthetically pleasing as the more traditional rear-mounting technique, it will do the job.

Hardware Selection

In this chapter, all nuts, bolts, screws, and other types of fasteners are referred to as hardware. This covers everything from the screws that fasten the communications equipment to the inside of the case, to the U-bolts used to secure the case to a pole or similar structure.

Hardware like the enclosures discussed in this section, come in many different varieties. This chapter deals only with Society of Automotive Engineers (SAE) hardware (bolts measured in inches). In addition, we only focus on plated steel and stainless steel hardware, as more exotic (and expensive) titanium, aluminum, and other "rare earth" alloy fasteners are beyond the scope of this book.

The hardware used in the construction of our outdoor networks can be broken down into four simple categories: bolts, nuts, washers, and screws.

Bolts

To the untrained eye, a bolt is just a bolt. It's made of metal and is threaded. You turn it right to tighten it and left to loosen it. This, however, is not the complete story. Just like the rest of the hardware discussed in this chapter, the first differentiating factor is the material the bolt is made from. By far the most common material is zinc-plated steel. These bolts are very shiny and highly resistant to rust and corrosion. Within this category, zinc-plated bolts can be further divided into grades. The only way to determine one grade of bolt from another is to inspect the head of the bolt where the wrench fits. The bolt grading scale ranges from grade 0 to grade 8, and though very rare, a grade 9 may be encountered from time to time. The higher the grade, the stronger and more expensive the bolt will be (see Figure 11.2).

Figure 11.2 SAE Grade Markings on Bolt Heads

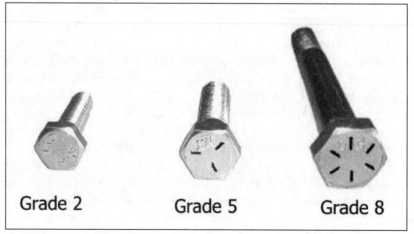

Grade 2   Grade 5   Grade 8

While a higher grade marking means a stronger bolt, it also means a more brittle bolt. Grade 2 bolts tend to bend when reaching the limits of their load-bearing capabilities. Grade 8 bolts, however, tend to shear with little warning when reaching the limits of their much higher load-bearing capabilities. For 99 percent of all applications associated with building outdoor wireless networks, a favorable balance can be found in grade 5 bolts, which are among the most common grades available and therefore competitively priced.

Plain steel bolts have no markings on their heads and are highly prone to rusting and corrosion; therefore, they should never be used in any outdoor applications. They rarely come in a grade higher than 2.

On the other hand, stainless steel bolts have the best rust and corrosion properties of any steel bolt. Often, an "SS" marking can be found on the bolt head indicating "Stainless Steel." If an "SS" marking cannot be found and you are unsure of the material, a simple way to determine if a bolt is made of stainless steel is to touch a magnet to its surface. If the bolt sticks to the magnet, it's not stainless steel, because one of the defining properties of stainless steel is the absence of iron. Iron is not only responsible for the magnetic properties of steel, but is also the element that causes plain steel bolts to rust. They do not, however, come in varying grades such as zinc-plated steel bolts and are substantially more expensive.

Regardless of the material of the bolt, the sizing fundamentals remain the same. All bolt sizes are expressed in the following format: 1/4-20 × 1, pronounced "quarter twenty by one." The "1/4" represents the diameter of the shaft, or shank of the bolt. This field is expressed in inches or fractions of inches. The "20" expresses the thread-pitch, which is an expression of the number of threads per inch. The higher the value of this field, the finer the thread. Finally, "1" represents the length of the shank of the bolt. This is measured from just under the head of the bolt to the end of the shank. This number is expressed in whole and fractional inches. Size increases of 1/8 inch increments are usually available from a 1/4-inch shank length all the way up to 1 inch. Between 1 and 3 inches they are typically available in 1/4-inch increments, and from 3 inches and up they are generally only available in 1/2-inch increments. Based on this knowledge, we can extrapolate from the format that a 1/4-20 × 1 bolt would be 1 inch long, 1/4 inch in diameter, and have 20 threads for every inch of threaded shank.

It is important to point out that the size of the head of the bolt where the wrench fits is not mentioned in the bolt-sizing format . The most important field in the format is the diameter of the shank. Our 1/4-20 × 1 bolt is referred to as a quarter inch bolt (the diameter of the shank), not by the size of the wrench needed to tighten it, which in this case would be a 7/16-inch wrench. The purpose of the format is to quickly and accurately describe the most important characteristics of a bolt to communicate how that bolt can be integrated into a project. Head size is rarely a deciding factor in this scenario.

For each size of bolt, there are two standard thread pitches: coarse and fine. However, the values differ between shank sizes. For instance, a coarse 1/4-inch bolt has a thread pitch of 20, while a fine 1/4-inch bolt also has a thread pitch of 20. A course 5/16-inch bolt has a thread pitch of 18, while a course 3/8-inch bolt has a thread pitch of 16. While this may seem confusing, after you determine the entire sizing format of the hardware you are using on your project it will become second nature.

Nuts

Nuts are also available in plain steel, stainless steel, and zinc-plated steel. Nut sizes are expressed in a format just like bolts. A nut that would fit a 1/4-20 × 1 bolt would be described as a 1/4-20 nut where "1/4" represents the diameter of the hole and "20" represents the thread pitch. It is important to note that the shank size and thread pitch have to match in order for a nut to thread properly to a bolt. Tightening a fine thread nut to a course thread bolt will result in cross-threading, which, in extreme cases, will ruin the threads of both the nut and the bolt. Nuts are also available in varying strength grades, although there are rarely markings of any type to indicate grade. Fortunately, the grade of a bolt is much more important than the grade of a nut.

Nuts can be further divided into two basic groups: locking and machine screw (see Figure 11.3.). Machine screw nuts thread onto a bolt of the same size and thread pitch with ease, until they reach an obstruction such as the surface the nut is intended to tighten to. In the locking category, there are a variety of locking mechanisms that prevent the bolt from being easily tightened or loosened to prevent the nut from working its way loose over time. The most common type of locking nut is known as a *nylock nut*. The locking mechanism of a nylock nut is a nylon ring crimped into the top of the nut that deforms around the threads of the bolt. Other locking nuts use mechanisms where the walls of the nut are deformed in a variety of different ways to prevent the nut from loosening on its own. These are collectively known as *crimp nuts*. Some form of locking nut is preferred in most outdoor wireless network applications, due to the fact that the equipment will probably be in a somewhat inaccessible location and routine maintenance will probably be out of the question, preventing the periodic tightening of fasteners.

Figure 11.3 Machine Screw Nuts and Various Locking Nuts

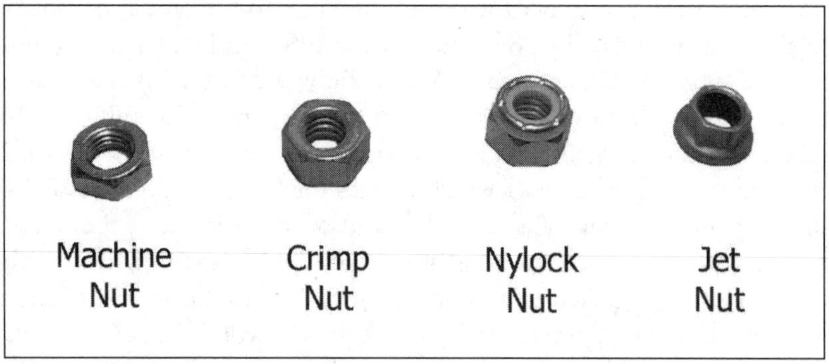

Machine Nut   Crimp Nut   Nylock Nut   Jet Nut

Washers

Washers are often overlooked components of fastener hardware because their function is not as obvious as that of nuts and bolts. Washers are available in all of the same materials as nuts and bolts and can be divided into two basic categories: flat washers and lock washers. A flat washer has two basic purposes. The first is to provide a smooth, even surface to allow the bolt head and nut to spread out the pressure generated over a larger surface. The second purpose of a flat washer is to protect the surface of the bolted object. Flat washers should almost always be used under the bolt head and under the nut.

Lock washers come in a variety of forms, but all are designed to keep a nut from loosening itself over time. The most common type of lock washer is a *split washer*. A split washer resembles an almost-closed "C"; one end of the C is bent up slightly from the other end. When a nut is tightened down on the split washer's surface, the pressure forces both ends of the washer into direct contact with the top and bottom surfaces. When counterclockwise rotational force is applied to the nut after it has been tightened down on a split washer, the portion of the washer that is bent up digs into the bottom of the nut, preventing it from backing off the bolt unless sufficient force is applied. Lock washers should always be used directly under the nut in conjunction with a flat washer.

Screws

The screw types discussed in this chapter are limited to machine screws, sheet metal screws, and wood screws.

Machine screws are just like bolts, only instead of having a hex head for a wrench they come with either a Phillips star head or a slotted head. Machine screws are measured in the same format as bolts only they usually start at sizes smaller than 1/4 inch, at which point the diameter is measured in screw size. Screw sizes are whole numbers, most commonly starting at 10 and working down in even numbers (i.e., 10,8,6,4, etc.). These sorts of screws are used to secure circuit boards to the inside of the enclosures.

Wood screws, like machine screws, have either a slotted head or a Phillips star head. However, unlike a machine screw, a wood screw has very coarse threads and a pointed tip to allow it to penetrate wood and thread itself through the grain. Wood screws are measured in standard screw sizes and thread pitch is sometimes included in this measurement.

Sheet metal screws are similar to wood screws in that they are coarsely threaded and pointed, although there are subtleties in the threads of each that allows them to be specially suited to their own categories. The biggest distinguishing feature between the two groups is that of the self-drilling sheet metal screw, which as the name implies, comes equipped with a crude drill head at the tip of the screw allowing for penetration without the need for drilling a pilot hole first.

The world of fasteners can be a bit intimidating because there are so many specialized areas, each with its own jargon. However, just like anything else, once the basics are learned the rest falls into place. If you are interested in learning more about fasteners, a few Web sites of interest are:

- www.atlanticfasteners.com
- www.mcmaster.com
- www.mscdirect.com

Performing the Hack

With a basic understanding of the materials at your disposal for constructing outdoor enclosures, we will now explore a few different methods for mounting sensitive electronic equipment in outdoor enclosures.

Metal NEMA 3 Enclosures

For this hack, you will use a PC Engines PC Wrap board and a steel NEMA 3 enclosure. First, you must determine the optimum orientation of the board inside the case. For this Wrap board, the choice is easy because all of the connectors are on one edge of the board. Because of the knockouts that come standard on the bottom of the case, the obvious choice is to place the board in the case so that all of the connectors are facing the knockouts.

Preparing the Case

Once the orientation of the board has been decided on, you must determine the location of the mounting holes for the board. One of the best ways to accomplish this is by creating a template.

To create a template, perform the following:

1. Place the board on a piece of paper and trace the perimeter of the board with a pen or pencil. Making sure that the board does not move on the paper, mark each of the mounting holes, as shown in Figure 11.4.

Figure 11.4 Tracing Board Perimeter and Mounting Holes

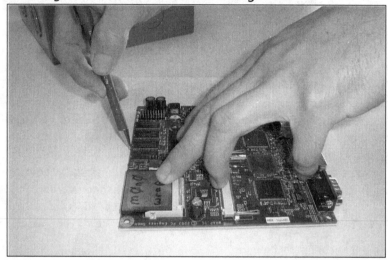

2. Now that the template is sketched, cut out the perimeter outline. At this point, it is a good idea to mark the template with "This Side Up" and with indicators as to which edge of the board has the connectors.

3. Place the template inside the case in the position in which the board will be permanently mounted.

4. Using scotch tape or something similar, secure the template to the case.

5. Using a punch and hammer, center the punch on the middle of the mounting hole outlines on your template and with a precise yet firm blow of the hammer, indent the steel underneath the template, as shown in Figure 11.5.

6. Repeat for the rest of the template hole outlines.

Figure 11.5 Using a Punch to Transfer Template Hole Outlines to Enclosure

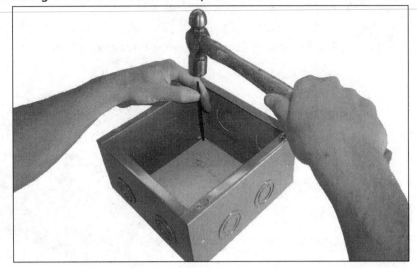

Now that the template marks are transferred to the back of the enclosure, you can easily drill holes in the exact locations of the mounting holes on the PC Wrap.

7. Carefully remove the template from the case (you may need it again later). Select the appropriate drill size (start with a drill bit that most closely fits the mounting holes of the board). One of the advantages of using the punch is the creation of an indentation in the steel that marks where to drill and also provides a center point for the tip of the drill to prevent wandering. To take advantage of this, begin drilling with the drill as close to a 90-degree angle as possible, using a slow rotational speed. Once enough material has been removed to create a large indention, you can increase the drill speed.

8. Sometimes the inner flange of the NEMA 3 case will create an obstruction, forcing the drill to an off angle when attempting to drill close to the edges of the enclosure. This will cause the drill to wander off center. If this is the case, leave those holes for last and concentrate on the holes you can easily drill. Once those holes are complete, turn the case upside down and affix your template to the back of the case, only this time with the side marked "This Side Up" facing down.

9. Next, align the template with the holes you drilled from the other side and secure with tape. As before, use the punch and hammer to transfer the marks on the template to the steel below.

10. Remove the template and drill.

11. Often, the drilling process leaves bits of metal protruding from the edge of the hole. This debris must be removed before you can proceed because it may cause problems later in the assembly process. The best way to remove the debris is to use a countersink bit on the drill,

as shown in Figure 11.6. However, if one is not available, it can often be removed by using a suitably sized Phillips head screwdriver as a de-burring tool.

Figure 11.6 Removing Burrs with a Countersink Drill Bit

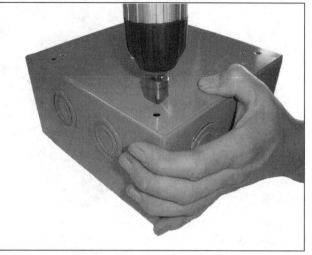

12. Now that the position of the board in the enclosure is finalized, remove the necessary knockouts at the bottom of the enclosure, as shown in Figure 11.7.

Warning: Hardware Harm

I would never recommend that these techniques be attempted while the actual equipment is in the enclosure. One slip of the hammer and the equipment that lay inside could be easily damaged. Creating a template requires extra work, but the effort is well worth it to protect your gear.

Figure 11.7 Using a Punch and Hammer to Remove Knockouts

Antenna Connectors and Bulkheads

With the board securely mounted to the standoffs in the NEMA 3 case and the appropriate knockouts removed for any necessary wires (Cat-5, serial, etc), we can now turn our attention to the antenna connectors. As discussed in Chapter 10, the most common antenna connector is the N-type, which is also the most robust structurally speaking and has the least RF loss of any other connector type. To complete this project, we are using a Senao mini Peripheral Component Interconnect (PCI) radio card with onboard u.fl connectors. The most conventional method of securing an antenna connector to an outdoor enclosure is to use a pigtail that terminates in an N-type female bulkhead. A bulkhead connector is a connector with a flange and machine jam nut to secure it to an enclosure wall.

Bulkhead connectors are typically "keyed," which means that it is oval in shape instead of round. When a keyed bulkhead connector is inserted into a keyed hole, the oval shape prevents the connector from spinning inside the hole.

Although it is acceptable to insert a keyed bulkhead connector into a round hole with a diameter equal to that of the keyed connector at its widest point, it does not provide any of the protection afforded by a keyed hole.

To create a keyed hole, perform the following:

1. Determine where on the enclosure you will mount the connector. It is always best to mount it on the bottom of the case whenever possible. When constructing enclosures of this type for the SoCalFreeNet, the bulkhead connectors are put all the way to the side of the bottom of the case to avoid all of the knockouts. Putting these connectors at the bottom of the case is also beneficial because placing them on the top or sides of the enclosure could lead to water and dust entering.

2. Determine the diameter of the N-female bulkhead at its narrowest point. This is usually 5/8 of an inch. If this is the case with your bulkhead, select a 5/8-inch drill bit and drill a hole in the desired location.

3. Using a dremel or similar tool, slightly egg out the hole to match the shape of the keyed bulkhead (see Figure 11.8). After a little bit of material is removed, check the size of the hole against the size of the bulkhead.

4. Repeat this process until it just barely fits. It is better to remove too little material rather than too much, as you can always remove more.

WARNING: PERSONAL INJURY

Be sure to wear gloves and safety glasses at all times when using a dremel or any other type of grinding tool.

Figure 11.8 Keying a Hole for an N-Type Female Bulkhead

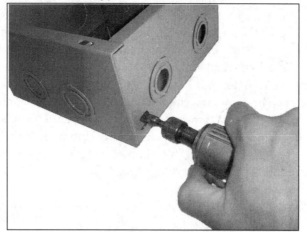

Mounting the Board

For this project, we will be using #6-32 aluminum standoffs to raise the board off of the back of the case. Aluminum was chosen for its low cost and anti–corrosion/rust properties.

Perform the following:

1. Secure the standoffs to the back of the case with #6-32 × 1/4" Phillips head machine screws (see Figure 11.9). A #6 screw is noticeably smaller in diameter than both the holes drilled in the case and the mounting holes in the board. This is to compensate for possible misalignments you may come across.

2. Lay the board in the case on top of the standoffs you just installed. With a little luck, the standoffs will line up perfectly with the PC Wrap board. If adjustments are necessary, loosen the screws on the back of the enclosure that are holding the misaligned standoffs in position to allow for play.

3. Once the board has been mounted from the top, tighten up the back screws again.

Figure 11.9 Mounting Standoffs to Enclosure

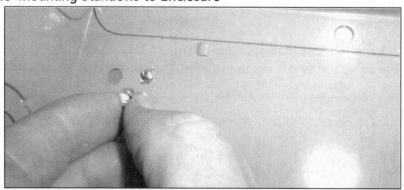

4. Sometimes one or two of the holes may be misaligned. If this is the case, use a dremel or similar tool to egg out the hole in the necessary direction. Remember to use a flat washer to cover up any area that the head of the screw does not cover.

WARNING: HARDWARE HARM

When mounting communications equipment in metal cases, particularly when that equipment is a bare Printed Circuit Board (PCB) , the risk of accidentally shorting out elements of the board is very real. Bolting the board directly to the back of the case is an extreme example of a poor installation, as all of the solder connections that protrude from the bottom of the board would be touching the back bottom of the metal case. This will not only cause the equipment to malfunction, but could permanently damage it. The best solution is to use standoffs to space the board off the back of the case. Standoffs are available in a variety of sizes and materials, from aluminum to nylon. Nylon standoffs are great to work with because they carry no risk of shorting out the board because nylon does not conduct electricity. Nylon standoffs can be harder to find and are usually more expensive.

Mounting the Enclosure

With the communications equipment securely mounted inside the outdoor enclosure, you can now begin to work on a mounting strategy for the enclosure. If the enclosure is going to be mounted to a wall, all you need to do is remove the board from the case, leaving the standoffs in place, and secure the enclosure to the wall with screws through the back of the case, either using the existing holes that come standard in NEMA 3 cases, or by drilling your own. Once the screws are in place and the case is secured, mount the board back onto the standoffs and secure the lid.

If you are planning on mounting the enclosure to a pole, however, things can get more complicated. The easiest and most conventional way of bolting an enclosure to a pole is to use U-bolts. The most important measurement of a U-bolt is the inner diameter (ID). If possible, this should correlate exactly with the outer diameter (OD) of the pole you will be mounting it to. The standard U-bolt ID sizes are: 1/4, 3/8, 1/2, 3/4, 1, 1-1/4, 1-1/2, 2, 2-1/2, 3, 3-1/2, 4, 5, and 6 inches. While it is possible to use a larger U-bolt than pole size, this should only be done when no other alternative can be found.

Once the pole size and the U-bolt size have been determined, you can focus on finding a mounting point on the enclosure. When mounting an enclosure on a pole, two U-bolts should always be used, one on top and one on bottom. This is to minimize the chance of movement, which overtime can cause fatigue and loosening of the U-bolt nuts.

Some NEMA 3 enclosures come with U-bolt flanges or "ears" on the outside. If this is the case, you're almost done. Unfortunately these flanges often come with holes for only one size of U-bolt. If the pole size you are bolting to is smaller than this, you can drill a new set of holes in the flange. If your NEMA 3 case doesn't have flanges, or if the pole you are bolting to is larger than the flange will accommodate, you will have to improvise.

As discussed earlier in this section, if you allow for a generous amount of space when sizing your enclosure you will have no problem drilling your own U-bolt holes in the back of the case above and below the communications equipment board.

When preparing to drill your own U-bolt holes inside the case, be sure to remove any electronic equipment that may be inside.

Perform the following:

1. Determine the clearance on the top and bottom of the board. If you have 2 inches of clearance top and bottom, turn the case over and measure half of that distance to find the middle. In this case, the exact middle is 1 inch from the top on the back of the enclosure, which will be in direct contact with the pole.

2. Mark 1 inch down on the left side and 1 inch down on the right side. Repeat the procedure for the bottom.

3. Using a straight edge, connect the two marks with a horizontal line. Do the same for the bottom.

4. Determine the vertical centerline of the enclosure. On the back of the enclosure where you marked out the clearances top and bottom, measure the width of the enclosure and divide by two. If the enclosure is 10 inches wide, make a mark at 5 inches at the top of the enclosure and at 5 inches at the bottom of the enclosure.

5. Using a straight edge, draw a vertical line connecting these two marks. This is the vertical centerline. Along the two horizontal lines where the vertical line intersects is where you will measure out from to drill the holes for the U-bolts (see Figure 11.10).

Figure 11.10 Locating Drill Points for Custom U-bolt Holes

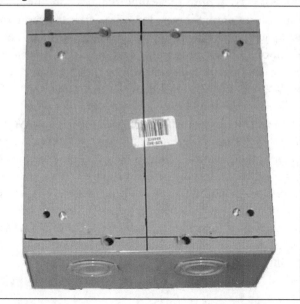

6. To determine the distance of the holes that are necessary to accommodate your U-bolt, you must first determine the distance from center to center of the threaded ends. For example, on a 3-1/4 inch U-bolt, that measurement is 3-7/8 inches. To properly space these holes, you must mark 1-15/16 inches (half the distance) on either side of the vertical line, along both horizontal lines. This will ensure an exact fit.

7. Use the hammer and punch to mark the four drill points and drill four holes of the proper diameter. If you do not have enough clearance to perform the preceding procedure because the board takes up too much room, you can modify the procedure to bolt up one of the other sides of the enclosure to the pole

With this final procedure complete, remount the equipment in the case, bolt the enclosure to the pole, and admire your handy work.

Under the Hood: How the Hack Works

The techniques reviewed in this chapter are only a drop in a very large bucket of methods and tools available for constructing enclosures. In your quest to shield your critical and expensive wireless equipment from the often brutal and unforgiving elements, don't be afraid to experiment or try something new. Even if it doesn't work out the way you wanted it too, chances are you'll learn something from the experience.

The common thread binding almost all of the techniques and procedures discussed in this section is that they are all based on standards. The Society of Automotive Engineers (SAE) standard and the National Electrical Manufacturers Association (NEMA) standard give us the ability to accurately articulate the size and strength of a bolt with a simple three-field format, or the level of protection offered by an enclosure with a single number. While it may seem overwhelming at first, learning to interpret these and other standards will make your projects much easier. A final word of advice: Don't be afraid to ask questions in a hardware or industrial supply store. Most of the time, the sales staff will be happy to help you learn.

Building Antenna Masts

Site surveys for new potential SoCalFreeNet nodes can be an interesting pursuit. First, all of the attention is directed towards determining the line of sight to certain key areas. Next, antenna selection comes into the conversation, followed shortly by talk about what sort of new piece of hardware or software should be attempted at the location. Then, invariably, someone asks a very sobering question: "What are we going to mount all this stuff on?"

This section will shed light on a subject that is often glazed over when preparing to deploy an outdoor wireless network, yet ironically, is the cornerstone of a successful wireless network. With this knowledge you will be better prepared to meet whatever rooftop mounting requirements come your way.

Preparing for the Hack

The most important thing to consider before embarking on an antenna mast building project is the type of roof it will reside on. For example, many SoCalFreeNet locations exist on the roofs of local residents who were kind enough to donate their rooftops to us. For this kind of application, it is beneficial to be able to offer a "passive" antenna mounting solution that does not penetrate the roof in any way. Other applications involve a roof so steep in pitch that mounting an antenna mast to it is near impossible, or at the very least, dangerous. In these situations, wall mount antenna masts may be the way to go.

Another item to consider is what sort of antenna you are considering for the antenna mast, and how many antennas will need to be mounted. For example, SoCalFreeNet employs several masts with multiple antennas. Many rooftops are deployed with an omni-directional antenna on an 802.11b/g frequency, and a directional antenna on an 802.11a frequency. Each of these antennas can have its own communications equipment housed in its own outdoor enclosure, and some can share the same communications equipment housed in a single outdoor enclosure. These sorts of issues should be considered before the design process of the antenna mast begins.

As obvious as this may sound, it is always a good idea to check for existing structures that can be used as antenna masts. Oftentimes, ready made antenna masts can be found with rusted, antiquated TV antennas clinging to them, from which no one has received a signal in years. This has happened on more than one occasion where a SoCalFreeNet team trying to formulate an antenna mounting strategy finally thought to ask the landlord if the TV antenna was used by anyone. We generally receive the reply, "What TV antenna?" and with permission from the property owner, we simply remove and discard the old antenna and install our new equipment.

As mentioned earlier, UV rays can be a destructive force in outdoor environments. Not only can they be damaging to enclosures and antenna mast materials, but they are equally damaging to Cat-5 wires and the zip ties used to secure them. Fortunately, both Cat-5 and zip ties are available in UV-resistant varieties. While they cost a bit more than the conventional types, their longevity in outdoor environments may be worth the cost, particularly in difficult to reach installations where just getting back up involves balancing precariously on ladders and climbing on steeply pitched roofs.

WARNING: HARDWARE HARM

Working on rooftops is very dangerous. Not only is there the danger of falling, but the very equipment you are installing can pose its own dangers as well. High voltage power lines often pass near rooftops. Caution must be taken not only to avoid contact between the power lines and the antenna masts while you are on the job, but also to locate the equipment in a position that will prevent contact with power lines and will keep the gear from falling off the building during high winds or structural failures.

NEED TO KNOW...A SAFETY NOTE

Always be sure to mount antennas and masts in such a way that even in the event of an accidental physical failure (or extreme winds), the antenna and/or mast would fall safely. Never mount equipment where it might fall onto power lines below.

Performing the Hack

After some thought has been given to what the roof topography looks like and what sort of antennas will be mounted to the structure, you can begin construction of the antenna mast. It is always best to begin a project like this with a good idea of what you are going to construct. While a detailed drawing is not necessary, rough measurements, the types and quantities of fasteners required, and the anticipated construction material needed are all variables that should be narrowed down as much as possible. This will save the time and headache of return trips to the hardware store for the one or two items you may have forgotten.

Equally as important is the work area where the mast will be constructed. A few minutes spent in preparation can save hours of frustration later in the project. Providing yourself with a clean and organized work area, anticipating and laying out all of the tools and supplies needed for the project, and generally getting all of your ducks in a row before you begin will go a long way in ensuring a smooth, successful project.

The Free-Standing Antenna Mast

The most versatile antenna mast design employed by the SoCalFreeNet is the free-standing antenna mast. This design does not require penetrating the roof or any other building facade with mounting hardware. Rather than obtaining stability through the anchoring of the structure to the roof, you derive the stability from a large weighted surface area. The following is the required list of tools for this hack (all of these items are readily available at any hardware store):

- Galvanized threaded steel pipe
- Galvanized threaded pipe flange
- Galvanized threaded pipe cap
- 1-inch thick 3-foot × 3-foot (or larger) plywood or similar
- 10 feet of wood 2 × 4
- 2-inch long wood screws
- 1/4 × 1-inch or larger lag bolts
- Sandbags or cinder blocks

The free-standing antenna mast consists of a wooden base 9 square feet in area, weighted by sand-bags or cinder blocks (or anything cheap and heavy) with a mast made of galvanized steel pipe, bolted to its surface.

Galvanized steel pipe makes an excellent choice, due to its easy availability, low cost, and high resistance to the elements. Almost all hardware stores carry this material, because it is a common item in everyday construction. The steel pipe most ideally suited to your application is threaded on both ends and comes in varying lengths and diameters. The common diameters are: 1/2, 3/4, 1, 1-1/4, 1 1/2, 2, 2-1/2 , 3, and 4 inches. Common lengths are: 48, 60, 72, and 126 inches. For our application, a 2-inch diameter and 72-inch (6 feet) length pipe will allow for plenty of strength and height. A threaded steel cap and flange of the same diameter can usually be found near the steel pipe in the hardware store (see Figure 11.11). The flange screws onto one end of the pipe and is bolted to the wood base. The steel cap is threaded on top of the steel pipe to prevent water from accumulating in the pipe.

Figure 11.11 Galvanized Steel Cap and Flange

To secure the mast to the base structure you will use lag bolts, which are very similar to wood screws, except instead of having a traditional screw head, they have the same type of hex head you find on a bolt, and they are available in much larger sizes than wood screws. Lag bolts are typically available in sizes from 1/4-inch to 3/4-inch, with lengths of up to 1 foot. For this application, however, you will be using 5/16 × 1-1/4 inch lag bolts. A 1-1/4 bolt length will compensate for the added material of the pipe flange, allowing the bolt to thread all the way into the wood base.

Once all of your tools have been organized, perform the following:

1. Determine the intended location of the flange before drilling the pilot holes. Pilot holes not only help guide the bolts into place making it easier to tighten them, they also help prevent the wood from splitting around the bolt by removing excess material that the bolt would otherwise displace outwards. For a 1/4-inch lag bolt, a 1/8-inch drill bit is recommended. For a 5/16-inch lag bolt a 9/32-inch drill bit should be used, and for a 3/8-inch bolt, a 3/16-inch drill bit should be used. For most applications, placing the steel pipe flange

directly in the middle of the base will be sufficient. This can be eyeballed, or you can use a centerline technique similar to that used to mark out the U-bolt holes in the previous section.

2. Use the holes of the steel pipe flange as a marking template and trace out at least four of the flange holes with a pen or pencil. This will allow you to drill without being encumbered by the flange ensuring a perfectly vertical pilot hole. In this case, you will be using a 9/32-inch drill bit to drill the pilot holes directly in the middle of the outlines, as shown in Figure 11.12.

3. After the holes are drilled, place the flange back in its intended location and check to make sure everything lines up. If by chance one or more of the pilot holes doesn't line up with the flange holes, rotate the flange 45 degrees, retrace the holes, and try again.

4. Once the holes line up, thread the lag bolts into the pilot holes and torque them down, being careful not to over-tighten them.

Figure 11.12 Drilling Pilot Holes for Lag Bolts

The wooden base of this antenna mast design provides for an inexpensive platform to mount the steel mast to, as well as provides a large surface area for the ballast. However, using wood does have a few drawbacks, such as poor longevity in areas with frequent rain. This can be overcome by using a wood-protecting water sealant or varnish that will protect the wood from water damage; however, it may need to be reapplied every few years. Plastic can be used instead of wood, although it is harder to find and more expensive. Ultra High Molecular Weight (UHMW) Polyethylene makes a fine material for this purpose; however, a 3-foot square, 1-inch thick sheet costs around $85.00. Again, www.mcmaster.com is one of the best places to research different materials such as UHMW or other plastics for use in this project.

This accounts for the basics of the free-standing antenna mast, but there is one more element of concern. Having a large flat object sitting on a roof for years is the perfect breeding ground for roof rot. This is due to water seeping under the antenna mast base and becoming trapped. This is the same as roof rot caused by leaves or pine needle accumulation on a roof. In some very dry environments, this is not a concern. Fortunately, for those of us who live in wetter environments, there is a simple solution.

5. Before the antenna mast is bolted to its base, cut a few lengths of 2 × 4 lumber to form a "plus" pattern on the bottom of the base, as shown in Figure 11.13.

6. Next, cut four more lengths of 2 × 4 and screw one to each of the four corners of the base. This will ensure the base's stability, while allowing water that may become trapped underneath to evaporate.

Figure 11.13 Elevating the Base to Prevent Roof Rot

7. With the base and the mast secured, you now need to acquire some ballast. As discussed earlier, all you need here is an inexpensive, dense material. Several free-standing antenna masts in SoCalFreeNet networks are weighted with sandbags. One of the biggest problems encountered so far with sandbags is the tendency of the actual bag to deteriorate over time. This can be avoided by re-bagging them in a more substantial material, such as UV-resistant plastic. Using bricks or cinder blocks is an acceptable solution as well, but be careful in wind prone environments. Either way, make sure you get at least 50 lbs. of ballast on your antenna base. All of the SoCalFreeNet free-standing antenna masts utilize at least 150 lbs. of ballast. This is probably overkill, but the piece of mind a few hundred pounds of ballast provides is well worth it. Figure 11.14 shows the finished product.

Figure 11.14 The Finished Product

Direct Mount Antenna Masts

The free-standing antenna mast is a great alternative when the antenna mast position has not yet been finalized, or when the property the mast is being installed on is not your own (the last thing you want is a call about a leaky roof caused by your antenna mast). However, there are times when you may want to bolt your mast directly to the roof. This can be a much easier solution compared to making a free-standing antenna mast base.

When considering a direct mount antenna mast, you must first survey the roof. If it is a perfectly flat or near perfectly flat roof, you need to look no further than the galvanized steel flange from the free-standing antenna mast design. Rather than bolting it to the base for a ballast platform, you simply bolt it directly to the roof. The process is the same, using the same lag bolts as before, but this time you must be sure to use roof sealant. After the pilot holes are drilled for the lag bolts, give each of the pilot holes a generous application of roof sealant, such as *DAP Watertight Roof Sealant*. Liberally coating the bottom of the flange mount will provide a further degree of protection. As the bolts are tightened, the roof sealant will be forced into and around the bolt holes, thereby providing a watertight seal.

Direct Mount Antenna Mast for Walls

The aforementioned methods work great when you have roof access, but what happens if you are unable to gain access to the necessary rooftop or the roof is too steeply pitched? Direct mount antenna masts for walls may be the answer you're looking for.

These antenna masts are also secured to the wall with a flange mounting system. Depending on the location, roof sealant may or may not be used.

For this portion of the hack, perform the following:

1. Determine the distance from the wall you would like the antenna mast to extend from. This distance will be equal to the length of the first tube you use. Screw or bolt it into the flange mount; it will protrude perpendicular to the wall.

2. Attach a 90-degree speed rail elbow fitting with set screws on the end of this pipe. The other end will accept another pipe of the same diameter. The vertical distance you want your antenna mast to clear from the wall mount will be equal to the length of this pipe. This pipe will also be secured with set screws.

Now that you have the basic "L" shape from the wall, you can mount your gear to the vertical pole. Some SoCalFreeNet installations use this type of mast for the antenna only, with the communications equipment mounted in an outdoor enclosure secured to the same wall, independent of the antenna mast itself.

If you have access to a metal shop, you can fabricate your own masts and avoid buying pipe fittings and/or costly speed rail fittings. As shown in Figure 11.15, a resourceful SoCalFreeNet member constructed a very inexpensive wall mount antenna mast using approximately $2.00 worth of scrap steel. To prevent rust, it was spray painted with industrial primer. Of course, this talented volunteer had a metal inert gas (MIG) welder, a drill a press, a sheet metal shear, and the expertise necessary to use them. These tools aren't necessary for creating quality antenna masts, but they give you more options.

Figure 11.15 An Inexpensive Wall Mount Antenna Mast Made From Scrap Steel

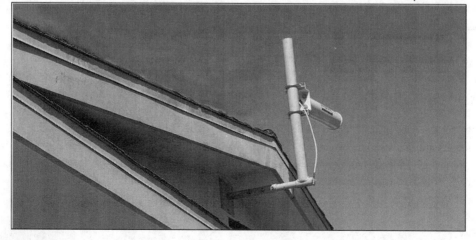

One of the advantages of this type of antenna mast is its ability to tilt vertically. This is accomplished by loosening the set screws on the horizontal side of the elbow, allowing the entire vertical assembly to rotate around the horizontal pipe. While not very useful for a simple omni-directional broadcast antenna, it can come in handy when attempting to align a directional backhaul antenna with a distant point either above or below the horizon requiring a subtle change in angle. In the case of the antenna mast shown in Figure 11.15, the backhaul link distance is just over a half a mile. Because the backhaul target is so far away, every degree of vertical tilt results in several tens of feet of change in the actual target point. This style of wall mount antenna mast is particularly useful for long backhauls because of its near infinite vertical tilt adjustability.

When vertical adjustability is not necessary, an L-shaped bracket can be used to secure an omni-directional antenna to an overhang without needing a traditional antenna mast. These brackets are commonly available at hardware stores, and are often referred to as *angle brackets* or *framing brackets*. This style of bracket is perfectly suited to this application because they cost almost nothing and are usually available galvanized. For added convenience, they come in several different styles with a variety of mounting holes. An ideal bracket will have one long and one shorter end. The longer end will be screwed or lag bolted to the overhang, and the shorter end will bolt to the antenna. Not all antennas are suitable for this kind of application. Many omni-directional antennas are only intended to be mounted to a traditional mast with proprietary clamps or standard U-bolts. For example, *www.hyper-linktech.com* carries a line of omni antennas that are intended to be mounted in this fashion. These antennas come with an N-type female bulkhead connector at their base. To take advantage of this bulkhead connector, you must drill a 5/8-inch hole (or slightly larger) in the L-shaped bracket. The female bulkhead on the base of the antenna will fit through this hole, and the supplied bulkhead jam nut will thread onto the connector. This will secure the antenna in the same fashion that the N-type female bulkhead connector was secured in the outdoor enclosure in the previous section. Then you must secure the bracket to the eave or overhang with wood screws.

Installing this type of antenna mast on an overhang is beneficial for a few reasons. First, raising the antenna up above the roofline allows for better signal propagation. Second, the wall below the overhang can provide a sheltered area to mount the outdoor enclosure that houses the communications equipment (see Figure 11.16).

Figure 11.16 L-shaped Bracket-mounted Antenna with an Eave-protected Enclosure

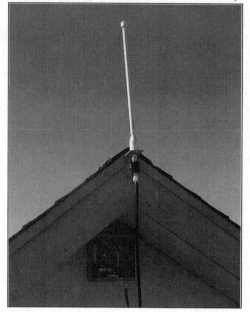

Lightning Protection

Regardless of the antenna mast type, it is always important to use a lightning arrestor whenever installing an antenna outdoors, particularly when that antenna is the highest point in the immediate vicinity. Not only is it a practical idea, but in some areas, local building codes require it. In addition, if you install the equipment on someone else's property, you could be held liable if a lightning strike to the equipment causes damage to their property. While installing a lightning arrestor does not guarantee to protect the equipment it is connected to or the property the equipment is mounted to, it does show that you took all of the appropriate precautions and made every effort to protect the property. This small investment can go a long way in protecting you from litigation if lightning causes damage to someone's property.

A lightning arrestor is a small device that is wired inline with the antenna and the antenna lead, as shown in Figure 11.17. The most common type of lightning arrestor for this application is known as a *gas discharge lightning arrestor*, which is composed of two major components. First is the gas discharge unit itself, and second is a ground shunt. When the difference of electrical potential between the antenna side of the arrestor and the antenna cable side of the arrestor reaches a high enough value (as occurs during a lightening strike), a change of state in the gas occurs, directing electrical conductivity to the ground shunt instead of to the antenna cable and thus protecting the rest of the system.

Figure 11.17 Lightening Arrestor in Place

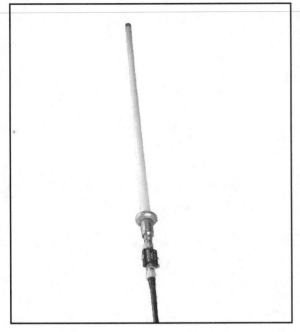

Because all electrical conductivity (and thus the lightning itself) is now being directed to the ground shunt, the shunt must now have continuity to ground. The most direct method of accomplishing this is to attach a grounding wire (un-insulated wire of a specified gauge) to the shunt, and run it all the way down to the ground. The wire must then be buried in at least 18 inches of earth.

In most homes and buildings, the third prong of a standard household 110-volt wall outlet, known as the *ground prong* or *lead*, is connected to a network of grounding wires throughout the building. This network of wires terminates in the same fashion as described, with wires plunging several feet into the earth. It is often easier to tap into this grounding network than to run your own lead directly into the earth. You may wish to consult with an electrician to determine the most effective way to accomplish these tasks. If you use a good prong, you must verify that the third prong is actually a true ground.

Summary

Building antenna masts is one discipline of building outdoor wireless networks where creativity knows no bounds. The procedures outlined in this section are designed to be accomplished with a minimal amount of equipment and expertise. For those that have access to more advanced equipment, the sky is the limit. Creating very tall guy-wired masts supporting arrays of antennas is not beyond the realm of possibilities. Many people may think this an extreme solution, as the time and cost is relatively high and the completed structure has a rather high profile. For the past 60 years, homeowners have been erecting huge antenna masts on their roofs, only with large unsightly TV antennas. If this kind of structure can be commonly erected for the purpose of watching four TV stations, then surely they can be easily erected for the purpose of building wireless networks. Following the guidelines in this chapter, you can begin building your own masts with relative ease and low cost.

## Chapter 12

Solar-Powered Access Points and Repeaters

Topics in this Chapter:

- **Constructing Solar-Powered Access Points and Repeaters**

Introduction

One of the most gratifying aspects of setting up wireless networks is bringing the network coverage to places where wires can't easily go, such as your backyard, a park, or even a distant building in your neighborhood. Unfortunately, places where wires can't reach are often places where grid-supplied electricity can't reach either. Detaching the network from the power grid is the last step in making a wireless network truly wireless and completely independent of its terrestrial components. Both battery technology and photovoltaic technology have come a long way since the days of the first solar-powered calculators. Modern solar technology has enabled a new era of truly independent wireless networks! In this chapter, you will learn how to:

- Calculate your power requirements
- Select the best batteries for your deployment
- Choose the right solar panel for your power needs
- Position your solar panel for maximum year round efficiency
- Build a rugged structure to support your equipment
- Wire your solar Access Point safely while minimizing the chances of failure

With this knowledge, we will then review in detail a real-world solar deployment, built by the members of the SoCalFreeNet Project in San Diego, California. We will learn what worked with this model, what didn't work, and how future solar deployments could be made even better. Finally, we will explore the possible applications of this exciting and extremely adaptable technology.

Preparing for the Hack

Before we can begin constructing our solar-powered Access Point, we must first take a look at the planning and research necessary to insure an orderly and productive construction experience. In preparation for this hack, we must coordinate the following items:

- Calculating power requirements
- Battery selection
- Selecting a solar panel

Calculating Power Requirements

The first thing we need to know about setting up a solar-powered AP or repeater is how much power the electronic gear will draw. Often times, this information will not be readily available in the literature that is provided with the equipment. Even if this information is available, it is often a "worst-case scenario" power draw. In other words, the power consumption the device is rated at is often much higher than actual or "typical" draw, as shown in Table 12.1. In such a situation, a little exploratory surgery can go a long way in effectively planning for a solar deployment.

Table 12.1 Popular Device Rated Power vs. Actual

Device	Rated Voltage	Rated Amperage	Calculated Wattage	Measured Wattage
Soekris Net 4521	12V	1.15A	14W	4.2W
PC Engine PC Wrap	12V	.4A	4.8W	3.12W
Linksys WRT54G	12V	1A	12W	11.4W

Let's assume for now that we don't know either the voltage or the amperage at which our equipment is rated. The vast majority of wireless gear available today runs on DC (direct current) power. Electricity for these devices is usually supplied by a small wall mounted transformer (also known as a power supply or "wall wort"), which plugs directly into a standard 110 volt alternating current (AC) wall receptacle. AC electricity entering the transformer is transformed (hence the name) into the DC electricity required by the equipment.

WARNING: HARDWARE HARM

The following procedure is not always necessary and could damage your equipment. While estimation of power consumption is generally considered to be acceptable, this procedure is used to obtain exact numbers to aid in planning for your power needs.

WARNING: PERSONAL INJURY

Working with electricity is an undertaking that should never be taken lightly. Always take precautionary measures such as unplugging devices from their power source before attempting the following procedures.

In this section, we will be focusing entirely on the DC side of the transformer. In order to measure power, we first have to get access to the copper underneath the wire's plastic insulation. First, as close to the transformer (AC outlet) side of the wires as possible, use a razor blade to separate the two wires from each other. Next, very carefully, make a small incision in each of the wires, just enough to get the probe of the digital multimeter (DMM) to touch the copper wires on each side. (See Figure 12.1.)

Figure 12.1 Testing Operational Voltage

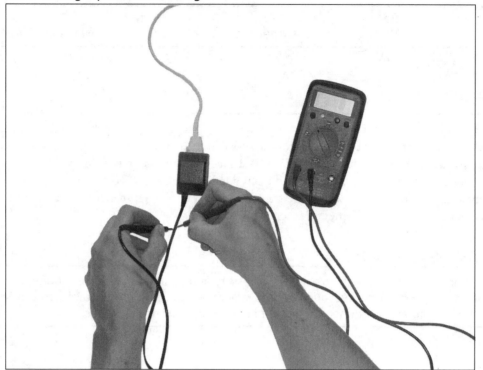

Next, plug the transformer in and plug the DC jack into the device and power it up. Wait until the device has finished its boot process, and then insert the DMM probes into the slits in the wires. Set your DMM to Volts DC and record the value. This is the device's voltage requirement.

Now that we know what voltage the device is operating at, the next step is to figure out how much current or amperage the device consumes. To do this, we need to configure our DMM to measure amperage. This usually involves removing the positive lead from the DMM and inserting it into a different jack on the unit. Refer to your DMM owner's manual for the correct procedure.

Unplug the transformer from the wall, and cut one of the wires where the incision was made. (Don't worry, we won't be using the transformer in the final product, and the wires will need to be cut anyway!) Set your DMM to measure amperage, plug the transformer into the wall, and put a probe on each end of the wire, as seen in Figure 12.2 (it doesn't matter which probe goes on which wire). Let the device finish its boot process and record the amperage.

Figure 12.2 Testing Device Amperage

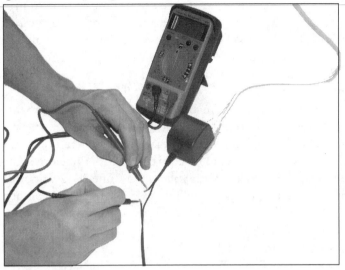

NEED TO KNOW...AC POWER

For some applications, alternating current or AC power may be required. In this case, a power inverter may be used. A power inverter takes the 12 volts DC from a battery or solar panel and turns it into 110 volts AC, the same as your household electrical current. It is possible to just run a power inverter and plug the wall transformers of your devices into that, as it saves the extra steps of cutting the transformer wires and hardwiring them directly into the batteries. However, it would be wise to avoid this since it not only introduces one more piece of equipment as a potential point of failure, but it's also one more device turning electricity into heat and wasting power, not to mention each individual transformer for your wireless gear creating it's own heat. Converting DC to AC to DC is extremely inefficient.

Now that we know the voltage and amperage, we can figure out how much overall power this device consumes, also known as watts. This can be easily calculated by the following formula:

Volts x Amps = Watts

Repeat this procedure for all the devices used in the solar-powered system.

Battery Selection

Now that we know how much power our devices will be drawing, it's time to determine our battery needs. All of the battery types we will be focusing on in this chapter will be lead acid. However, these are not the same kind of battery you would find under the hood of your car. That type of battery is not designed to be discharged much beyond 80 percent of capacity. Instead, deep cycle lead acid batteries, as their name implies, can go much deeper into the charge/discharge cycle without damaging the battery. Deep cycle batteries can be further divided into two main categories: flooded and gel cell. The flooded type contains water and must be checked from time to time to make sure the cells are completely immersed in H_2O. They do typically offer more energy storage capacity then the gel cells and cost a bit less. On the other hand, gel cells require no maintenance at all. Therefore, they are sometimes referred to as "maintenance free" deep cycle batteries. For this reason, I would highly recommend the use of gel cells in all solar deployments.

Deep cycle batteries typically come in 6- and 12-volt flavors and a whole variety of storage capacities, referred to as amp hours, or Ah. An amp hour is a way to measure the storage capacity of a battery and represents the number of amps of electrical current that the battery can provide in a one-hour period. Another way to think of amp hours is that they represent the number of hours that a 1-amp current-drawing device will be powered by a particular battery before it runs out of energy. Most of the equipment we will be dealing with will run just fine at 12 volts, thanks to their internal voltage regulators. For this reason, this chapter will focus on multiple 12-volt batteries wired in parallel. (See Figure 12.3.)

Figure 12.3 Batteries Wired in Parallel

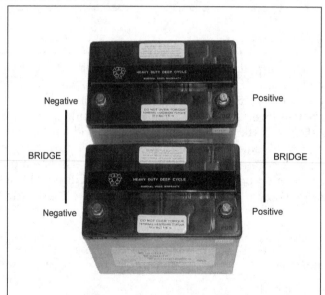

This is where all that work to determine the power requirements of the equipment comes in handy! Lets say we will be using two 12 volt batteries, each with 50 Ah of storage wired in parallel. Keep in mind that batteries wired in parallel double their storage capacity, while the voltage remains constant. With the previous formula of Volts x Amps = Watts, we just plug in the battery values:

12 x (2 x 50) = 1200

This means that our hypothetical setup will give us 1200 watt hours of energy. If these batteries were going to power a device that our test showed required 7 volts at .3 amps, then our formula would show:

7 x .3 = 2.1 watts

Dividing 1200 watt hours by 2.1 watts gives us 571 hours of runtime!! That's almost 3.5 weeks of power, and we haven't even factored in the extra energy provided by the solar panel yet!

Selecting a Solar Panel

We've determined what kind of hardware we will be running in the system, and planned out how much battery capacity we'll need. Now we need to decide what kind of solar panel we want.

Just like batteries, solar panels are rated with voltage and amperage values; however, they are usually measured by overall wattage. Choosing the right solar panel for the job brings us back to our previous formula: Volts x Amps = Watts. This time, however, there are more factors to take into consideration. For example, a node in Seattle will require a larger solar panel than an identical node in Palm Springs. This is the result of both cloud cover and latitude. (See Figure 12.4.)

Figure 12.4 Sun Hours by Region

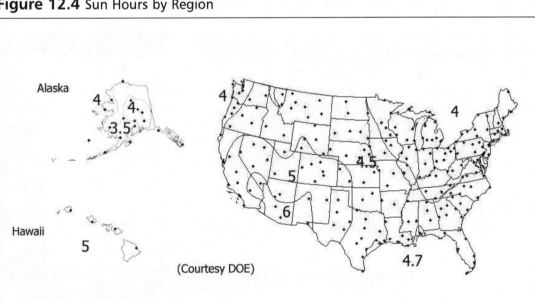

Image courtesy of U.S. Department of Energy

The fact that cloud cover is a variable in the performance of a solar panel is not particularly surprising to most people. However, latitude as a performance factor can be a difficult concept to grasp immediately. For example, Table 12.2 shows a selection of ten geographically diverse cities along with their high, low, and average sun hours.

Table 12.2 High, Low, and Average Daily Sun Hours by City

City	High	Low	Average
Los Angeles, CA	6.14	5.03	5.62
Seattle, WA	4.83	1.60	3.57
Miami, FL	6.26	5.05	5.62
New York City, NY	4.97	3.03	4.08
Cleveland, OH	4.79	2.69	3.94
San Antonio, TX	5.88	4.65	5.30
New Orleans, LA	5.71	3.63	4.92
Bismark, ND	5.48	3.97	5.01
Lexington, KY	5.97	3.60	4.94
Fairbanks, AK	5.87	2.12	3.99

The part of the Earth most ideally suited to solar deployments is the equator, where latitude is 0 degrees. At the equator, the days are longest and the sun's angle is the most direct. The further you move from the equator, either north or south, the more indirect the sun's angle and the fewer hours of usable light per day. This is due to a decreasing angle of incidence. Figure 12.5 demonstrates that as the angle of incidence decreases from 90 degrees to 45 degrees, the surface area covered by the same sunlight increases and the amount of solar radiation per square inch decreases. In other words, the sunlight gets "diluted" across a wider swath of surface area on the solar panel. This decreasing angle of incidence as latitude increases has a direct effect on usable sunlight. For example, in San Diego in June, the average sun hours total about 6.5 a day, but Maine only gets about 4.5 sun hours a day in the same month. To determine the sun hours in the area where your solar node will operate, there is an excellent table of sun hours of cities in the U.S. at www.bigfrogmountain.com/ sunhoursperday.cfm. If you can't find your city in the list, the closest city to you will do fine.

Figure 12.5 Angle of Incidence

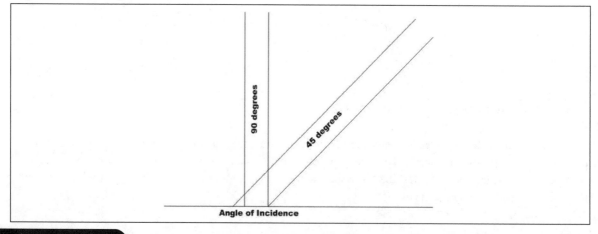

Solar panels can be purchased in a myriad of different wattages, from 1 to 180 watts (and even more, should a special need arise). Using the average sun hours rating in the area you plan to deploy your solar node, it's easy to determine the size of the panel required to fit your needs. For an average of 6 hours of sun a day, a 55-watt solar panel will produce approximately 330 watt hours. By dividing 330 watt hours by our 12 volts, we arrive at 27.5 amp hours. This would be enough power to keep our previously mentioned solar node running for a very long time, with plenty of overhead to account for adding more equipment down the road, long patches of bad weather with little sun, and device inefficiencies.

Using these calculations, selecting the proper battery/solar panel combination can be a mathematical certainty, instead of being based on a hunch or a guess. If you don't trust your own math, or would just like a second opinion, there are some great calculators online at www.bigfrogmountain.com/calculators.cfm.

Now that we know the basics of a solar-powered Access Point, it's time to get down to the nuts and bolts of building it. We're going to learn how it all comes together to harness the power of the sun and provide coverage in that critical area in a wireless network beyond the reach of power lines. The most important elements of the project are as follows:

- Solar panel

- Batteries

- Charge controller

- Wires and wiring components

- Structural materials

- Wireless gear

- Power inverter (optional)

Everything outlined here is readily available in any major metropolitan area, with the exception of the solar panel itself and the charge controller. This is not to say that there won't be a photovoltaic solar panel distributor in your area, but they may be harder to find. The following web sites are a great place to start your search and to develop an accurate perspective of the industry:

- www.solar-electric.com

- www.siliconsolar.com

- www.altenergystore.com

- www.solar4power.com

- www.bigfrogmountain.com

These sites not only have a great selection of solar panels, but also offer everything you need to install, wire, and secure your solar panel.

Most of these online stores carry deep cycle lead acid batteries as well, but shipping these can be quite expensive. In some cases, the shipping can cost as much as the battery itself! In most cities, you will have no problem finding stores that carry deep cycle gel cells at reasonable prices. Recreational vehicle (RV) accessory stores are a great place to find a decent selection and knowledgeable sales staff who will help direct you to the best battery for your application. Looking in the yellow pages often leads to battery specialty shops, which can also be a good place to find deep cycle batteries. Specialty shops have a selection that can't be beat, but this convenience often comes at a premium in cost.

The solar panel itself is waterproof, but the batteries, charge controller, and power inverter (if one is used) should be housed in a rain-resistant enclosure. Home improvement stores like Home Depot are a great place to find enclosures of all types. See Chapter 11 for more information about all kinds of outdoor enclosures.

Building the supporting structure for your own solar-powered node can be as complicated or as simple as you want it to be. As long as you have a good mounting point for your solar panel, a place for your batteries and other electrical equipment (shielded from the elements), an enclosure for your network gear (preferably separate), and a good mast for your antenna(s), you're good to go!

Determining the angle of your solar panel, relative to the ground, is the first step in the design of your structure. There are volumes of information on solar web sites, with all kinds of advice as to how to angle your antenna to best take advantage of the position of the sun. Some even go so far as to use elaborate multiaxis trackers which keep the solar panel at the optimum angle for the season and time of day! These systems, however, are beyond the scope of this book, as the purpose of our solar-powered nodes is to provide low-cost, ultra-reliable network connectivity.

This doesn't mean that we're just going to toss a panel out in the sun and hope it gets enough sunlight either. There are some steps we can take which cost nothing to implement and will ensure that you're taking full advantage of the available sun in your area. The first step is to point your panel to true south (or north if you live south of the equator). This will position the panel in such a way that the sun will never be behind it (imagine the sun arcing from east to west over your southerly positioned panel). The next step is to determine your latitude. This measurement is expressed in degrees and has a direct correlation to the degree at which you position your panel. Most Global Positioning System (GPS) units will display your exact location in coordinates of longitude and latitude. Simply power up your GPS unit in the intended location of your solar node, and record the latitude reading. If you don't own a GPS unit, a useful Internet mapping utility can be found at www.maporama.com. It's convenient and works worldwide. Simply enter your street address, follow the prompts, and you will find a map of your city, with your exact coordinates in longitude and latitude at the bottom of the map.

Now that we're prepared with all of this information, we can determine the exact angle of the panel. This is easier said than done since the optimum angle changes with the seasons. During the summer months, the sun approaches its most direct angle, while during the winter months, it approaches its most indirect angle. For example, if you were to stand outside at noon in July and measure the length of your shadow, you would find that standing in the same place in January at noon would yield a much longer shadow. Now instead of your shadow, imagine the shadow of the solar panel. Position the panel due south at noon in the summer time and try to create a shadow as close to the actual size of the panel as you can. This is accomplished by placing the panel at a 90 degree angle to the sun. Come back in January and you will find the shadow stretched way out of proportion compared to the actual size of the panel.

Of course, we can be more scientific than simply looking at, and chasing, our shadows. Surprisingly, the methods for determining the optimum angle of photovoltaic solar panels can vary depending upon whom you ask. Unfortunately, there is no one "right" way, but the commonest school of thought in determining the angle of the solar panel is to take your longitude in degrees and subtract 15 degrees from that angle for the summer and add 15 degrees for the winter. Another school of thought says to multiply your longitude by .9, and then add 29 degrees for winter. With this method, you will arrive at an angle several degrees steeper than expected. This is to compensate for the concentration of solar energy during the winter months at and around noon. Of course, others will recommend that you just place the panel at your exact longitude and let it all average out. I will leave this up to you to decide what is best for your particular application.

In San Diego, for example, we are at a latitude of 32.7473 degrees. Adding 15 degrees to this gives us an optimum angle of 47.7473 degrees for the winter months, while subtracting 15 degrees for the summer months gives us an optimum angle of 17.7473 degrees. When deploying the San Diego solar node, it had been determined that the node was so over-engineered, in terms of battery capacity and solar generation capacity, that the additional calculations were not necessary and it was just set up at 32 degrees. At this angle, we felt confident that it would work fine, given the average daily sun hours in San Diego.

When setting up your solar node, be sure to take into account subtle things such as rooftop pitch or hilltop pitch, whichever the case may be. A very simple yet effective method for determining the exact angle of an uneven surface is to use a bubble level to represent the horizontal plane. Then use a protractor to find the true angle, as shown in Figure 12.6.

Figure 12.6 Finding the True Angle Using a Bubble Level

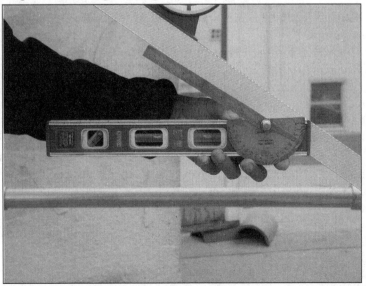

Now that we've covered the basics of what goes into a successful solar-powered Access Point or repeater, let's discuss the solar-powered repeater built for the SoCalFreeNet Project. This node was built not out of the necessity for a grid-independent node, but as a test bed for future repeaters on hilltops where the luxury of grid supplied electricity did not exist. After all, if something goes wrong, it's much better to test it out on a nice flat rooftop than at the top of some remote mountain! At the time of this writing, the node has been operational for about six months with absolutely zero failures. In the next section, we will review in detail the procedures and equipment that made this solar-powered repeater such a success.

Performing the Hack

Now that we have an understanding of what goes into the planning of a solar-powered Access Point, we can now get down to the specifics of the construction phase of this hack. The following section of this chapter outlines the construction materials and techniques implemented during the construction of the SoCalFreeNet solar node. The methods outlined here should not be thought of as the only way to achieve your objective, but rather as an insight into one of the many possibilities for achieving your objectives.

Structure

For the solar-powered repeater built for the SoCalFreeNet project, we started with a very substantial and expensive structure. Several four-foot and eight-foot lengths of 1-5/8" aluminum tubing were purchased from a local metal supply shop. They were all joined together with special purpose-built aluminum "Speed Rail" elbows, "T" pieces, and feet. (See Figure 12.7.) Aluminum was chosen for this project because this particular building owner was very concerned (and rightfully so!) about the structure's aesthetics. Aluminum, being a non-ferrous metal has a high resistance to the elements, ensuring an attractive appearance for many years to come. This was a rather expensive method, but since cost was no object for this particular property owner, it was a great way to build a structure that will endure for years. Whatever building material you choose, keep in mind it may have to support up to 100 lbs. of batteries or more!

Of course, not everyone is fortunate enough to have elaborate project funding. Luckily, for those on a more conservative budget, there are still many options. Galvanized steel, (the same type of material used for chain link fence posts) is an excellent alternative since it is inexpensive and stands up to the elements just as well as aluminum. Spray painting regular steel with a primer coat is another option. Though not as well protected from the elements as galvanized steel or aluminum, it still provides a degree of rust protection, though care must be taken to avoid scratching the painted surface and thus exposing the bare steel to moisture. Finally, pressure-treated lumber can also serve as an excellent outdoor building material. It is, of course, not as easily manipulated as the other materials, but will stand up to the elements for many years.

Figure 12.7 All-Aluminum "Speed Rail" Fittings

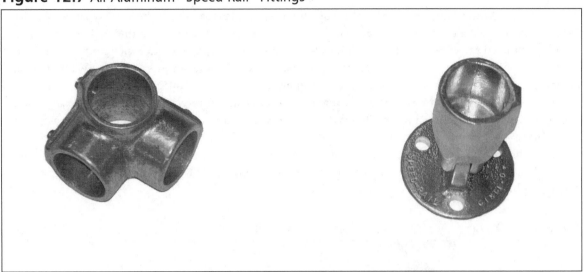

Using aluminum "speed rail" fittings resulted in a construction method that was remarkably easy to set up since each tube simply slid into the aluminum elbow or "T" piece and was locked into place with a series of set screws. (See Figure 12.8.) This was especially handy when it became apparent that the structure would not fit through the door that led to the roof once the structure was completely set up! And so, the structure had to be broken down to its basic components and hauled up several flights of stairs to the roof one piece at a time. Another advantage of this construction method is that the basic shape of the structure is a cube. Therefore, most of the aluminum tubes were exactly the same length. This allowed the structure to be torn down and re-erected in a matter of minutes.

Figure 12.8 The All-Aluminum Structure

Solar Panel

The solar panel mounting structure of the SoCalFreeNet node was built to be somewhat adjustable. (See Figure 12.9.) Since the raw latitude of San Diego was used to determine the angle of the panel, it isn't optimized for winter or summer. This also means that power generation per hour of sunlight will not experience any significant variation between the winter and summer months. While the calculations prove the ability of the solar panel to generate adequate amounts of power at this inclination and location throughout the year, the solar panel mounting structure was built to be easily adjustable should the panel's angle need to be altered during the winter months. This adjustability was achieved by securing the bottom of the panel to a fixed rail of the node's structure and the top of the panel to a rail capable of sliding up and down two of the structure's uprights. With this configuration, adjusting the angle consists of loosening a few set screws on the aluminum "T" pieces and sliding the bar up or down until the desired angle is achieved.

Figure 12.9 Vertically Adjustable Solar Support Rail

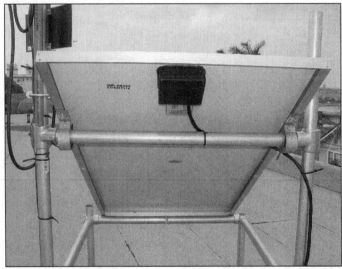

Electrical

For a battery box, a decision was made to use one of the in-ground utility boxes often seen housing city sewer and water control valves in urban neighborhoods. (See Figure 12.10.) The reason for using a utility box was to build the most indestructible solar-powered repeater possible. Needless to say, this was not the most cost-effective option, as the box itself cost around $200! But you can bet it's not going to break anytime soon! For the cost-conscious, any enclosure to shelter the electrical equipment from the elements is acceptable. Common household items such as large Rubbermaid containers or storage bins could serve this function.

WARNING: PERSONAL INJURY

Lead acid batteries are capable of producing extraordinarily high amperage. As soon as you introduce the batteries into a working environment, great care must be taken to avoid directly shorting the positive and negative terminals together. This can happen directly at the terminals, or indirectly at the terminal block. A direct short of the batteries could result in explosion, wire meltdown, and personal injury. Please exercise great caution at all times.

Figure 12.10 In-Ground Utility Enclosure

Inside the utility box sits two 6-volt deep cycle gel cell lead acid batteries wired in series. (See Figure 12.11.) The bridge between the positive and negative terminals of the batteries was made with 2 American Wire Gauge (AWG) battery wire. Once again, this is overkill in the extreme, but wire is not very expensive, and there is no harm caused by wire being too thick in distances this short. However, I would not recommend going any smaller than 12 AWG.

Figure. 12.11 Batteries Wired in a Series

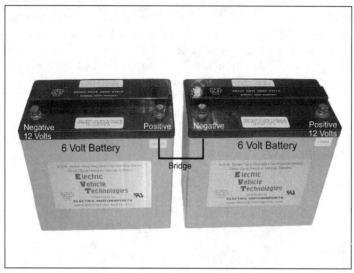

The details of the internal electrical wiring are actually quite simple. With the exception of the externally mounted solar panel and communications equipment, nearly all of the wiring is housed completely in the in-ground utility enclosure. This keeps the elements out, and keeps the electrical cable distances as short as possible. This is important due to the nature of electricity. The greater the length of wire used, the more resistance the transmitted voltage must overcome. The greater the resistance, the more heat builds up in the wires. The hotter the wires get, the higher the resistance becomes, and so on and so forth. For this reason, it is very important to run the proper wire gauge or AWG for the power and distance required. (See Table 12.3.)

Table 12.3 Recommended Wire Gauges at Specified Wattage and Distance

Distance	50W	75W	100W	150W	200W	300W
0–10 ft	12 AWG	12 AWG	12 AWG	12 AWG	12 AWG	12 AWG
10–20 ft	12 AWG	12 AWG	12 AWG	10 AWG	10 AWG	10 AWG
20–30 ft	12 AWG	12 AWG	10 AWG	8 AWG	8 AWG	8 AWG
30–40 ft	12 AWG	10 AWG	8 AWG	8 AWG	6 AWG	6 AWG
40–50 ft	10 AWG	10 AWG	8 AWG	6 AWG	4 AWG	4 AWG
50–60 ft	10 AWG	8 AWG	8 AWG	6 AWG	4 AWG	2 AWG

As you can see in Table 12.3, as distance and overall power increase, using an adequate wire gauge becomes very important. If, for example, you decided to put the solar panel of the roof, but house all the equipment in a room a few stories down, choosing the right gauge of wire becomes critical. Even if you decided to go with a design similar to that of the node outlined here, I would advise against running wires any smaller than 12 AWG on the main power transmission equipment. This includes

- Bridging the batteries (either in series or parallel)
- From the batteries to the terminal block
- From the solar panel to the charge controller
- From the charge controller to the terminal block

For a design such as the one outlined here using 12 AWG wire may seem overkill, but I assure you it is not. When designing such a system, it is always prudent to hope for the best, but plan for the worst. For example, if for any reason the solar panel were to stop charging the batteries, over time, the batteries would eventually completely discharge and the node would fail. After inspecting the solar system and resolving whatever the problem was, the panel would begin to recharge the batteries again—only this time, it's a brilliant summer day, and your 60-watt solar panel would be cranking out power as fast as it could, and the batteries, being completely drained, are happy to take the energy. In such a scenario, lesser gauge wires would begin to build up heat very quickly. As the heat built up, the resistance would increase causing even more heat. Before you know it, the wires have begun to melt-down and start a fire. I'll leave the rest of this scenario to your imagination. As always, please use extreme care with any electrical system. To help determine the right gauge of wire for the job, see Figure 12.12 for a side-by-side comparison of different gauges of wire.

Figure 12.12 Assorted Gauges of Wire

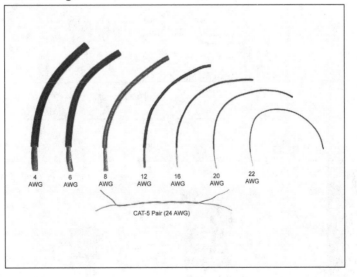

For the SoCalFreeNet node, a six-terminal block was used. The 12V supply from the batteries is fed to the first two terminals via 12 AWG wire, terminating in ring connectors. (See Figure 12.13.) As the batteries in this node are wired in series (the positive terminal of one battery is bridged to the negative terminal of the other—see Figure 12.11), the first terminal of the block is connected to the remaining positive battery terminal, while the last terminal on the block is connected to the remaining negative battery terminal. (See Figure 12.14.) Attached to the next inboard block terminals are the positive and negative leads from the solar charge controller. Finally, the DC transformer from the Soekris was cut, and two ring connectors are crimped to the wires in its place. These ring connectors are secured to the next set of inboard screws on the terminal block. Finally, all three positive terminals are daisy-chained together (see Figure 12.15) and all the negative terminals the same.

Figure 12.13 Wiring Diagram for a SoCalFreeNet Solar Repeater

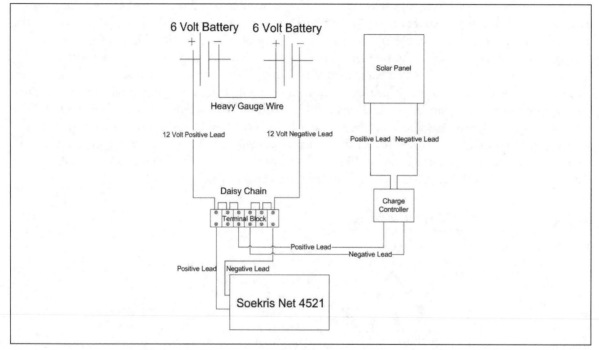

Figure 12.14 Terminal Blocks

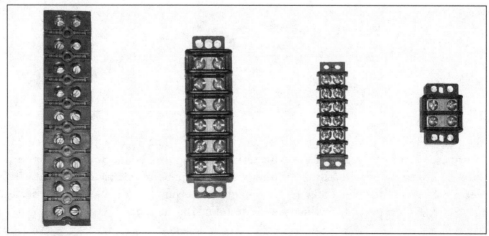

Terminal blocks are preferred over other means of securing electrical wiring since changes can be easily made and all the wire junctions are laid out before you, making troubleshooting or adding devices in the future much easier. For each terminal on the block, there are two screws, one on top and one on the bottom. This is for connecting multiple wires together, as each pair of terminal screws conduct electricity between each other. It is acceptable to attach one- or two-ring connectors to each terminal screw. If more positions are needed, you can daisy-chain multiple terminals together, as shown in Figure 12.15.

Figure 12.15 Daisy-Chaining a Terminal Block

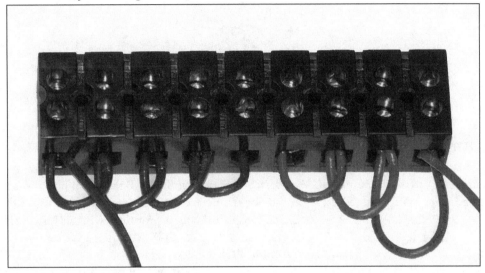

The charge controller is responsible for not only feeding current from the solar panel, but also preventing current from the batteries back feeding to the solar panel and destroying it. The charge controller itself has two sets of wires. (See Figure 12.16.) Two wires go to the solar panel, and two wires go to the batteries via the terminal block. It is a rather straightforward procedure since these devices are very simple and very rugged. In fact, during the initial testing of the SoCalFreeNet solar node, while positioning the unit on the roof, one of the SoCalFreeNet volunteers inadvertently placed his hand on the charge controller and found it to be extremely warm. It seems that during the reassembly process (following the teardown to get it up on the roof), the wires to the battery side of the controller had been reversed—positive to negative and negative to positive! It must have been like this for nearly an hour! The problem was quickly corrected and the unit has gone on to provide many months of flawless, uninterrupted service.

Figure 12.16 A Charge Controller with Two Pairs of Wires

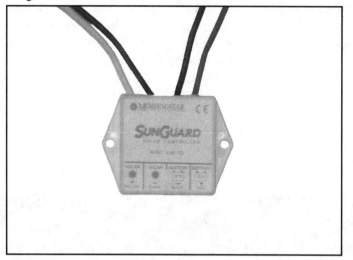

Electronics

The radio communications equipment in this solar node consists of a Soekris net4511. With support for one mini PCI card and one PCMCIA card, it is the perfect Single Board Computer to set up a repeater. See Chapter 4 for more details on the Soekris device. For the backhaul radio, an Atheros-powered 802.11a PCMCIA card was used. This card came with a proprietary pigtail that terminates in a standard SMA connector. For the Access Point radio, a Senao 200 mW mini PCI card was used. This card uses the u.fl connector common to mini PCI wireless devices.

The Soekris net4511 and wireless cards are housed in a standard NEMA 3 steel enclosure. See Chapter 11 for more information on outdoor enclosures. Both pigtails terminate in an N-female bulkhead which protrudes from the bottom of the case.

To secure the case to the structure, two 2" zinc-plated steel U-bolts were used. Because the U-bolts protrude into the case and the Soekris board has a footprint nearly equal to the length of the bottom of the case, it was decided that for this installation it would be easier to mount the case to the pole on one of its side walls. This was somewhat of a departure from the more traditional method of securing a case from the bottom wall, but it still resulted in a very sound and structurally protected deployment.

The same structural beam that the case is attached to also provides a mounting base for the antennas. A patch panel is used for the 5.8 GHz 802.11a backhaul antenna. This provides not only a highly directional radiation pattern, but also does so in a compact and inexpensive form factor. This antenna is mounted using the standard mounting hardware issued with the antenna. Two galvanized steel flanges are secured to the antenna with four small Phillips head screws. These flanges are, in turn, secured to the structure with the two stainless steel hose clamps provided by the antenna manufacturer. This allows for easy antenna realignment with a few turns of a screwdriver. While this does provide for rotational movement, antenna pitch (that is pointing the antenna up or down) is not inherently available in this design. Fortunately, this antenna's particular radiation pattern specified a very focused horizontal beam width while allowing for a rather generous vertical beam width. This creates an antenna which returns very pronounced signal intensity changes when adjusted from side to side, yet hardly noticeable changes when tilted up or down.

The AP antenna is also mounted to the same pole as the backhaul antenna. This antenna, however, did not come with any mounting equipment. Nevertheless, another two zinc-plated steel 2" U-bolts were used to secure this antenna to the top of the pole very nicely. (See Figure 12.17.) Mounting two active antennas this close to one another is usually frowned upon; however, it is quite acceptable to do so in this situation. This is possible due to the fact that 802.11b and 802.11a have massive frequency separation. This allows for the two antennas to be placed nearly on top of each other without interfering with one another. This is also the reason why two wireless cards with different frequencies can be operated in the same enclosure. When creating a repeater with two 802.11b radios, placing the radios and antennas at least 10' away from one another is highly recommended.

Figure 12.17 2.4-GHz and 5.8-GHz Antennas Mounted to a Structure

Under the Hood: How the Hack Works

Now that we have the batteries, the networking gear, and the solar panel in place, we are left to ponder the magic of a wireless node free from the dependencies of the electrical grid. In this section, we will take an in-depth look at what makes this setup work. Understanding how the hack works under the hood is not only fascinating, it's also a valuable tool for troubleshooting any unexpected problems which may crop up in the years of service you can expect from your solar node.

The Batteries

Common sense would lead us to believe that batteries store electricity; however, in reality, batteries store the *potential* for electrical generation. In other words, the electrical generation capacity within the battery is actually the result of a chemical reaction. If you were to crack open a lead acid battery (which is not recommended under any circumstances!) you would find sets of alternating lead plates pressed back to back with plates made of lead dioxide, immersed in sulfuric acid. These sets can be stacked upon one another in any quantity. The more sets of plates (also known as cells) that are stacked together, the higher the voltage between the first plate and last plate. This combination of dissimilar metals and electrolytes is known as a voltaic pile, named after its inventor, Alessandro Volta (Volta built the world's first battery in 1800, well before mechanical electrical generators were invented). The typical 12-volt lead acid battery has six such cells, each producing approximately 2 volts. Because they are stacked in series, the total voltage is approximately 12 volts. This is actually where a battery gets its name from, as it is a battery of cells! The type of battery you would put in a flashlight, such as a D cell or AA cell, aren't actually batteries at all. Those batteries only have one cell, hence the term D cell.

The positive and negative terminals of our deep-cycle batteries are connected directly to the first plate of the first cell and the last plate of the last cell. Between these two plates, the battery has its maximum potential. If you were to bridge these two plates with a conductor such as a metal wrench or copper wire (which is extremely dangerous and should NEVER be attempted!), the full potential would be realized and electrons would be flowing freely from the positive side to the negative side of the battery. This will eventually result in either a complete meltdown of the battery, an explosion, or both! This is known as a short circuit or dead short and should be avoided at all costs. Instead, a load must be placed between the positive and negative terminals. To simplify, we'll use the example of a light bulb.

The light bulb acts as a current limiter or resistor, allowing only a small amount of current or amperage to cross the two terminals of the battery. All of the equipment in our solar node acts in the same way as the light bulb: bridging the positive side of the battery to the negative side, thus completing the circuit, while only allowing a small amount of current to flow through the circuit.

The Solar Panel

Once again it may surprise you to learn that like the battery, the principles that make the photovoltaic solar panel work were discovered long ago. In 1839, the photovoltaic effect was first observed by the French Physicist Antoine-César Becquerel. His simple experiment proved that voltage could be produced by exposing an electrolyte solution to sunlight.

Photovoltaics remained little more than a scientific curiosity until the 1950s when the U.S. government began heavily funding photovoltaic research in an effort to power spacecraft indefinitely, in response to the Russian advances in space technology.

Output efficiency from these solar cells remained at about 1 percent throughout most of the 1950s. It wasn't until the late 1950s when efficiency began to approach 10 percent and solar cells were finally deployed in space aboard the American satellite Vanguard. Modern solar efficiencies are approaching 15 percent, although depending on the quality and type of solar cell, that figure can be as low as 8 percent.

Today's most cost-effective solar cells are known as thin-film solar cells. They get this name from the manufacturing process of applying super-thin layers of semi conducting material on a solid backing. This semiconductor material is very similar to the material found in computer chips such as Intel's Pentium line. The layers are so thin that they only measure about 1 micron or .001 millimeters together. The layers are made up of a positive and a negative layer of silicon, separated by the "intrinsic layer." As sunlight enters the intrinsic layer, it generates electrons. The positive and negative layers generate an electric field which forces the electrons through conductors and collect as they pass through more and more cells, finally exiting through the wires and into charge controller.

If you are interested in learning more about solar power, I would recommend the following books:

- *Practical Photovoltaics: Electricity from Solar Cells* by Richard J. Komp (an excellent book with lots of background and technical information)

- *The Easy Guide to Solar Electric* by Adi Pieper (focuses more on the practical aspects of implementing solar cells in the home)

- *Ugly's Electrical References* by George V. Hart (one of the best electrical reference books out there; you will find it to be an invaluable resource)

Appendix A

Wireless 802.11 Hacks

Topics in this Chapter:

Introduction

Hacking wireless hardware is an endeavor steeped in a rich history of experimentation and hobbyist culture. The wireless hardware hacker of today pursues his or her craft with a passion not seen since the amateur radio (also known as "ham radio" or "hams") operators of the last generation. Many wireless enthusiasts are, in fact, connected with the ham community. Once solely the domain of a small group of Radio Frequency (RF) engineers, wireless gear has never been so inexpensive and accessible as it is today. With rapidly declining hardware costs, anybody can learn and experiment with 802.11 equipment with only a small investment.

In this chapter, we review several wireless hacks, tricks, and hardware modifications, including:

- **D-Link DWL650** Card modification for adding an external antenna.

- **OpenAP (Instant802)** Reprogramming your Access Point (AP) to run an open-source version of Linux.

- **Dell 1184 Access Point** Exploring the embedded Linux operating system.

WARNING: PERSONAL INJURY

Please use extreme care when performing any kind of experimentation with RF devices. For more information about the dangers of RF exposure, visit the following URLs:

- www.wlana.org/learn/health.htm
- www.arrl.org/tis/info/rfexpose.html

NEED TO KNOW...

802.11 is a protocol created by the Institute of Electrical and Electronics Engineers (IEEE). This protocol defines a method for transmitting and receiving data communications wirelessly. The original specification was ratified in 1997. This protocol supported 3 physical methods: Frequency Hopping Spread Spectrum (FHSS) and Direct Sequence Spread Spectrum (DSSS) in the 2.4GHz frequency range, as well as Infrared (IR). (Note that IR was never successfully deployed as a commercial option). Speeds of 1Mbps and 2Mbps were supported. In 1999, the IEEE approved 2 new higher speed additions to the protocol: 802.11a and 802.11b. 802.11a defined (up to) 54Mbps Orthogonal Frequency Division Multiplexing (OFDM) at 5GHz and 802.11b defined 5.5Mbps and 11Mbps using DSSS in the 2.4GHz spectrum. In 2003, 802.11g was established to provide (up to) 54Mbps OFDM in the 2.4GHz spectrum. For more information about the 802.11 protocol, please visit http://grouper.ieee.org/groups/802/11/.

Wireless NIC/PCMCIA Card Modifications: Adding an External Antenna Connector

Wireless Network Interface Cards (NICs) typically have a PC Card (also referred to as PCMCIA) form-factor for use in laptops. These cards come in two basic varieties:

- Those with external antenna adapters
- Those without external antenna adapters

For example, Cisco AIR-PCM35x cards have integrated diversity dipole antennas, while the Cisco AIR-LMC35x cards have dual MMCX connectors (no antenna is supplied with the device). Figure A.1 shows a Cisco card with an integrated antenna, while Figure A.2 shows a Cisco card with dual MMCX connectors.

Figure A.1 A Cisco Card with an Integrated Antenna

Figure A.2 A Cisco Card with Dual MMCX Connectors

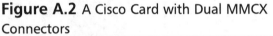

For typical indoor applications, an integrated antenna should work just fine. However, these antennas are often low on gain (2.2dBi) and lack the range needed for long distance applications. Having an external antenna connector is desirable because it gives you flexibility. This is particularly important for hobbyist applications (such as connecting Pringles can antennas). When it comes to setting up networks and experimenting with wireless LANs, having flexibility is a key benefit.

Historically speaking, cards with external antenna adapters were sold at a premium compared to cards with integrated antennas. This meant that hobbyists had to either cough up additional cash for a more expensive card or hack their own solution using off-the-shelf parts. Can you guess which path we're going to take?

Preparing for the Hack

In this hack, we will be modifying a D-Link DWL-650. Note that a number of variations exist for the D-Link 650 card so ensure you obtain a standard 16-bit PCMCIA PC Card by comparing your card to the one in Figure A.3. This is important because D-Link sold 32-bit CardBus NICs for a short time and called them D-Link 650! So really, the only way to be absolutely certain that the card you have on hand is the correct one for this hack is to look at the card...

The items you will need for this hack are:

Figure A.3 An Unmodified D-Link DWL-650 Card

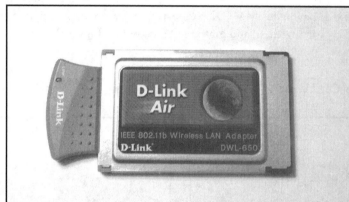

- **D-Link DWL-650 Wireless NIC** An inexpensive PCMCIA card that lacks an external antenna.

- **BNC cable** A small length of Thinnet cable connected to a BNC connector.

- **Soldering iron** To solder the connector wires to the leads on the Printed Circuit Board (PCB).

- **X-ACTO knife** To create a hole in the plastic casing for the wiring.

- **Tweezers or a toothpick** To open the casing and gain access to the PCB.

WARNING: HARDWARE HARM

This hardware modification will void your warranty and potentially violate FCC regulations. This hack should be used for test purposes only and not in a production environment.

You can choose from a variety of cable connectors for attaching your card to an external antenna. Some vendors have proprietary connectors, such as the Cisco MMCX connector. Other connectors are industry standard, such as BNC and N. In our hack, we will use a BNC connector. Don't worry if your antenna uses something other than BNC, because adapters to convert BNC to pretty much any kind of connector are easily available and inexpensive. For more information about RF Connectors, the following Web sites are useful:

- www.rfconnector.com
- www.therfc.com

Furthermore, it's fairly uncomplicated to find a BNC plug with a short length of cable, since you can simply take any old BNC Thinnet cable and snip off the connector (saving a few inches of cable). Figure A.4 shows an unmodified Thinnet cable.

Figure A.4 An Unmodified Thinnet Cable

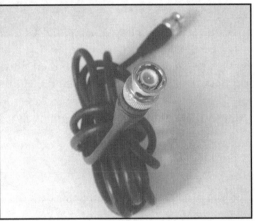

Performing the Hack

Hacking the DWL-650 to add external antenna functionality is performed in three basic steps:

1. Removing the cover.
2. Desoldering a capacitor and soldering it to an adjacent spot on the PCB.
3. Soldering the BNC connector's leads to the PCB.

Removing the Cover

The first step is to open the top cover of the D-Link DWL-650 card and expose the internal antennas. To do this, turn the card upside down (with the MAC address label facing you) and get your tweezers handy. On the short gray tab (protruding from the silver PCMCIA card), you will see four small clasps that hold the plastic cover in place as denoted in Figure A.5. Each clasp contains two semicircular plastic posts. Use your tweezers to squeeze the semicircular plastic posts together. You might also use a small flat-head screwdriver to wedge the gray plastic pieces apart as you advance to each plastic post. If you don't have tweezers, a toothpick can be wedged into the outer circle and used to leverage the semicircle posts toward each other.

Figure A.5 The Semicircular Posts Holding the Card Together

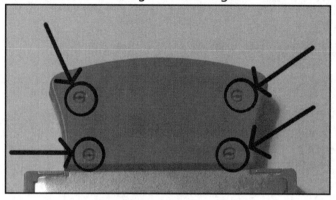

Once the posts are sufficiently squeezed together, you can then slowly and gently pry the gray cover off. Note that the cover is held in place by two gray tabs tucked neatly into the silver PCMCIA card edges. You may need to use a small flat-head screwdriver to carefully pry the silver chassis open just wide enough to slide out one end of the gray tabs. Be particularly careful here, since it's nearly impossible to fix the edges of the silver chassis once they are bent by mistake. Figure A.6 shows the cover once it's been removed.

Figure A.6 The Gray Cover

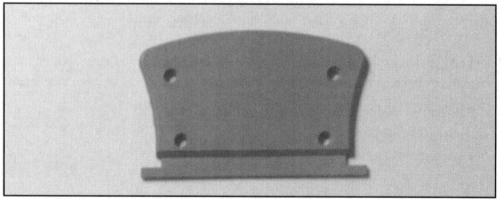

After exposing the antenna compartment (Figure A.7), you should see some silk-screened labels next to two surface-mount capacitors (C144 and C145). These capacitors are connected to the card's dual internal antennas. You will also notice a silk-screened label for ANT3. That's right—it seems as if

the D-Link card was actually designed to have a third, external antenna. However, it was never implemented and this connector lays dormant—until now!

Figure A.7 Inside the Antenna Compartment

Moving back to your gray casing and the gray cover you just removed, you will need to bore out a small hole to make room for the BNC cable to pass through. Take your X-ACTO knife and carefully scrape out a small hole in both the top and bottom portions of the gray casing as shown in Figure A.8. Note that this task is best performed by first removing the gray cover (as described in the previous step) and then separately boring out a semicircle in each half of the gray casing.

Figure A.8 Making a Hole in the Gray Casing

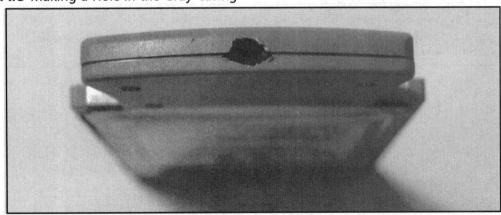

Moving the Capacitor

The second step of the hack is to desolder one of the surface-mount capacitors, either C144 or C145. It doesn't really matter which one you choose to remove. Once the capacitor has been removed, you need to resolder it to the lead labeled ANT3. You can do this by rotating the capacitor 90 degrees. If you chose C145, you will need to rotate the capacitor 90 degrees clockwise. If you chose C144, you will need to rotate the capacitor 90 degrees counterclockwise. In effect, all you are doing is disconnecting one of the surface-mount antennas (the thin diamond-looking leads on the PCB) and electrically connecting it to the ANT3 leads. Figure A.9 shows a close-up of the antenna compartment before the hack.

Figure A.9 Antenna Compartment: Close-Up Before the Hack

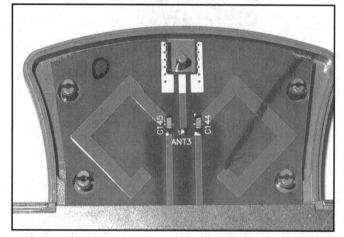

WARNING: HARDWARE HARM

Be very careful when applying heat to the circuit board with your soldering iron. Too much heat may damage the capacitors or the board and may cause the pads and traces to lift up, causing irreparable damage. In practice, it helps to use the tweezers to hold the capacitor in place in the new location while you carefully apply the solder. Sometimes it helps to have a second person involved to lend a hand with the tricky procedure.

Attaching the New Connector

The final step in the D-Link hack is to connect your BNC adapter to the PCB. Prepare your connector by removing the plastic covering and shielding to expose the center conductor and the surrounding copper strands. Figure A.10 shows a prepared BNC connector.

Figure A.10 A Prepared BNC Connector Ready for the Hack

The BNC connector should be soldered to the PCB at the gold-colored leads near the top of the antenna compartment. The center conductor of the BNC connector needs to be soldered to the center lead of ANT3, while the outer strands should be soldered to the adjacent pads on the PCB.

Once the soldering is complete, you can reattach the gray plastic cover by sliding in one of the little tabs and wedging the rest of the cover back in place. You should be able to snap it back together with ease. If there is too much resistance, go back and get your X-ACTO knife to widen the hole for your BNC cable a little more. Your completed hack should resemble the image in Figure A.11.

Remember to be very gentle with your new external BNC cable, because the force of inserting and removing antennas may cause the solder connections to come loose from the internal PCB. Many hobbyists have resorted to applying electrical tape or other reinforcing methods to keep the cable in place and protect the connections from getting damaged. One popular method is to use a hot-glue gun to apply glue to the cable and the surrounding gray casing.

Figure A.11 The Finished Hack: D-Link DWL-650 with External Antenna Connector

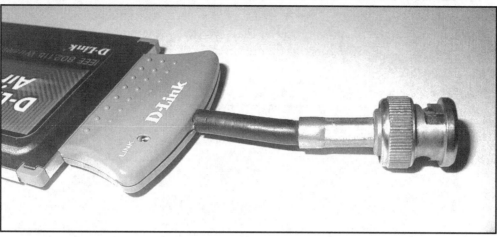

Under the Hood: How the Hack Works

This D-Link hack is a simple example of how you can gain additional functionality from a product that has been intentionally unused by its vendor. It is clear from the PCB design that this NIC was originally intended to support an external antenna (as seen by the ANT-3 pads on the PCB), but it was never implemented. Moving the capacitor simply disables part of the internal antenna circuitry and makes the signal available to an external connection. Note that your card will still work even if you don't attach an external antenna, because diversity mode will allow the still-connected second antenna (on the PCB) to transmit and receive a useable signal.

Diversity mode works by monitoring each antenna and automatically switching to the antenna with the stronger signal. This mode helps reduce errors caused by multipath problems. Multipath errors occur when signals bounce off objects in the transmission path and the same signal arrives at the receiver two times.

OpenAP (Instant802): Reprogramming Your Access Point with Linux

Wireless Internet access using standard off-the-shelf APs is lots of fun. However, traditional consumer-grade APs can be quite feature limited. Sure, it's possible to take an old PC, run Linux, and build your own AP, but the hardware form factors for old PCs tend to be large, clunky, and noisy. With a small form factor, you can install your AP in hard-to-reach places, weatherproof boxes, or tucked away in a corner. Wouldn't it be cool if we could take an off-the-shelf AP and reprogram it with a Linux operating system? That's what you can do with OpenAP, a free software package from Instant802 (http://opensource.instant802.com).

Preparing for the Hack

OpenAP is a completely free and open-source software package. However, it works only with certain hardware devices. Specifically, you need an AP that is based on an Eumitcom WL11000SA-N chipset. The good news is that these can be found in a number of consumer-grade APs, including:

- U.S. Robotics USR 2450 (Figure A.12)
- SMC EZConnect 2652W
- Addtron AWS–100
- Netcomm NP2000AP

Figure A.12 The U.S. Robotics USR 2450

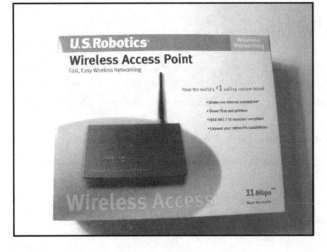

These devices can often be found at aftermarket resellers and on eBay and other online sources. Along with the appropriate AP, you also need the following items:

- **Linearly-mapped external memory card** This card needs to be 2MB (minimum) and operate at 3.3 volts. The OpenAP Web site (http://opensource.instant802.com) recommends a MagicRAM Industrial SRAM Memory Card. Another option for the memory card is the Pretec FA2002.

- ■ **Null modem cable** This is the cable you will need to connect your AP to your computer. Using the console port and a terminal emulation program (such as the Windows-based "HyperTerm"), you can communicate directly with your AP.

This hack does not require any special tools, however be sure to have the following items on hand when performing this hack:

- ■ **Screwdriver** To remove the screws of the AP box.

- ■ **9/16" Wrench or Pair of Pliers** To remove the antenna connection from the external box.

- ■ **Needle Nose Pliers** To remove the metal bracket holding down the PCMCIA card.

Performing the Hack

The first step in performing the hack is to obtain the source code for OpenAP which can be downloaded from http://opensource.instant802.com/sources.php. As of this printing, the most current version of OpenAP is 0.1.1. You also need to get the Linux kernel source and untar it into the OpenAP directory. The application source code and kernel source code URLs are provided in the following command-lines.

You can follow these commands to compile the Flash image:

```
wget http://opensource.instant802.com/downloads/openap-0.1.1.tar.gz
tar -xzvf openap-0.1.1.tar.gz
cd openap-0.1.1/
wget http://ftp.kernel.org/pub/linux/kernel/v2.4/linux-2.4.17.tar.gz
tar -xzvf linux-2.4.17.tar.gz
patch -p0 < ./misc/openap-linux-2.4.17.patch
```

By typing **make** and pressing **Enter**, you will be presented with the *makefile* options, as follows:

```
[root@Stephanie openap-0.1.1]# make

Makefile for OpenAP tools, kernel and flash image.


targets -
  tools      : build uclibc and assorted tools
  install    : install uclibc toolchain (must be root)
  bootstrap : configure and build kernel, then flash
  sram       : make sram image


Please see Makefile for details.
[root@Stephanie openap-0.1.1]#
```

In the openap-0.1.1 directory, take a look at the README file for more configuration details. Alternatively, if you prefer to download the prebuilt image from OpenAP rather than compiling your own, you can download it from http://opensource.instant802.com/downloads/sram.img.

Keep in mind that the image file size is 2MB. In order for OpenAP to work, the size of the image file must be equal to the maximum capacity of your SRAM card. This means that a 4MB card must have a 4MB image file. Once you've created the .IMG file (or downloaded it from the OpenAP Web site), you must adjust the file size to the matching size of your card. In a DOS environment (for a 4MB card), you would type the following command: **copy /B sram.img+sram.img sram2.img**.

After your .IMG file is ready, you need to copy it to your SRAM card. In a Windows environment, you can use a program called Memory Card Explorer to transfer the file. You can download a 30-day evaluation version of the Memory Card Explorer software application from www.syn-chrotech.com/products/software_02.html.

In a Linux environment, you can do both steps (doubling the file size to 4MB and copying the file to the SRAM card) with the following command (assuming that Linux identified your device as /dev/mem0c0c):

```
cat sram sram > /dev/mem0c0c
```

If you had a 2MB card, the command would simply be:

```
cat sram > /dev/mem0c0c
```

Now that your SRAM card is ready, the next step is to install the card into the AP. The idea behind using the SRAM card is that you will boot off of the card only one time in order to program the AP's Flash memory with Linux. After that, all future upgrades can be performed remotely, without reinserting the SRAM card.

Installing the SRAM Card

To install the SRAM card, start by opening your AP case. In our example, we will be using the U.S. Robotics USR 2450. Figure A.13 shows the AP before modification.

Figure A.13 The USR 2450 Access Point before Modification

Before you can remove the screws, you need to remove the antenna. This is a simple RP-TNC connector which can be unscrewed in a counterclockwise direction. The RP-TNC connector is held in place by a large nut at its base. Remove this nut with a 9/16" wrench or a pair of pliers. Once the antenna has been removed, you can remove the four screws on the bottom of the AP. Next, gently slide the cover off. Figure A.14 shows the AP with the cover removed.

Figure A.14 The AP with the Cover Removed

With the plastic cover out of the way, you will see the wireless NIC in the PCMCIA slot. The card is protected and held in place by a metal brace (Figure A.15).

Figure A.15 Before Removing the Card … See the Metal Brace?

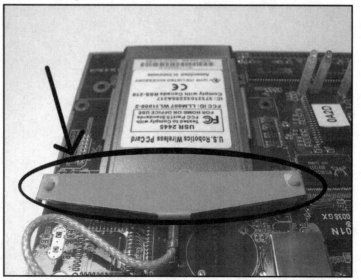

The metal brace can be removed by squeezing one of the plastic posts with a pair of pliers. These plastic posts are similar to the "old school" motherboard spacers that are used to mount computer motherboards to a chassis. Be careful removing these posts, because they can break easily. We recommend removing the post located near the edge of the PCB, as you can use a small string (running between the PCB hole and the hole in the metal bracket) to replace to post in the unlikely event of accidental post damage.

With the metal bracket dislodged, you can now remove the wireless card and set it aside. You'll reinstall it later, so keep it handy.

Next, install the SRAM card in the PCMCIA slot previously occupied by the wireless card.

Before you do anything else, you need to locate the JP2 jumper (two pads surrounded by a white box and "JP2" printed on the board). With the PCMCIA edge connector facing away from you, the JP2 jumper is located below the group of three LEDs, directly to the right of the Flash chip and to the left of the CPU, as denoted in Figure A.16. When JP2 is shorted (connected) on power-up, the device will boot from the SRAM memory card in the PCMCIA slot instead of booting from the on-board Flash memory. Using a paper clip or short piece of wire, connect the two pads of JP2 together as shown in Figure A.16.

Figure A.16 Close-up of the JP2 jumper, with a paperclip inserted

The Flash image you built and installed on the SRAM card contains code to cause the Access Point to write the OpenAP firmware into the AP's Flash memory, so you won't need to reperform this step. Future upgrades can be performed via the OpenAP software.

Power Me Up, Scotty!

The final step of the hack is to power up the device with JP2 shorted and the SRAM card installed. You should observe a green LED and a yellow LED flashing alternately. Once you see this flashing, release the JP2 jumper by removing your paper clip or wire while keeping the device powered up. Be careful not to touch any other components with the paperclip or wire. When the install from SRAM is complete, the device will reboot itself and (assuming the JP2 short is removed), will boot directly from its on-board Flash. You can observe this process by noting that the green and yellow LEDs will now flash back and forth more quickly. This process can take several minutes, so be patient.

Once the process is complete and the device has rebooted, you can remove power and reassemble the AP by reinserting the Wireless NIC, fastening the metal brace, screwing on the top plastic housing, and reconnecting the antenna connector to the outside of the case. With the AP reassembled, you can now fire up a laptop with your favorite AP discovery tool (such as NetStumbler, dStumbler, or Kismet) and look for an AP called instant802_debug (a list of popular AP sniffer tools and their URLs can be found at the conclusion of this chapter). If you see this SSID, your upgrade is successful and you now have a fully functional OpenAP device running Linux!

Under the Hood: How the Hack Works

Under the hood, the U.S. Robotics USR 2450 Access Point is basically a low-powered, single-board computer. It has an AMD ELAN SC400 CPU (based on the Intel 486 core) with 1MB of Flash ROM, 4MB of DRAM, and an RTL8019 NE2000-compatible Ethernet Interface IC. Connected via the PCMCIA interface is a Prism2-chipset Wireless NIC. OpenAP is an elegant hack that essentially takes advantage of this known hardware configuration and replaces the operating system on the AP with its own firmware.

One of OpenAP's most exciting features is the fact that it supports 802.11d bridging. This means that you can expand your wireless network and repeat your wireless signals across several "hops." Most APs connect to the Internet using a wired cable into a digital subscriber line (DSL) or cable modem, but OpenAP can get a connection to the Internet via an adjacent OpenAP (or another 802.11d compliant device) and then extend the coverage of the cloud to anybody within range of its own signal. In essence, your OpenAP can serve as both an 802.11 client and an 802.11 access point at the same time! A group in Palo Alto, California, has developed a cooperative community wireless network built around the OpenAP platform. Visit www.collegeterrace.net for more information about this exciting grassroots movement.

Having Fun with the Dell 1184 Access Point

Following the initial release of the Dell 1184 Access Point, rumors swirled about the possibility that Dell had used an embedded form of the Linux operating system in the device. An exhaustive port scan of the entire port range reveals the following open ports: 80, 333, 1863, 1864, 4443, 5190, and 5566. (A port scan is a search for open ports on a particular host.) Port 80 is for the standard built-in

Web server. This Web server is used for client configuration via a browser. The first clue that Linux is running can be found in the banner information for port 80, which reveals:

```
Server: thttpd/2.04 10aug98
```

Further exploration reveals that a Telnet daemon is running on port 333. This is where the real fun begins. Although this hack might not void your warranty and is more software-based than hardware-based, the fact that Dell exposes a Telnet service on an AP running a Linux distribution translates into hours of exploration and fun for all of us hardware hackers. Linux is a free and open source operating system with unlimited capabilities. In this section, we will explore the inner workings of the Dell 1184 through its open Linux services.

Preparing for the Hack

Preparation for this hack is simple. All you need is a Dell 1184 and another computer to Telnet into it from the wired or wireless segment. Figure A.17 shows a Dell 1184 Access Point.

Figure A.17 A Dell 1184 Access Point

NEED TO KNOW... THE DELL 2300 802.11G ACCESS POINT

There are various models of Dell Access Points. While the Dell 1184 AP supports 802.11b, the Dell 2300 AP supports 802.11g. Note that the Dell 1184 AP hack does not work on the Dell 2300 AP, so be sure you obtain the proper model before beginning the hack. These devices appear physically similar; however, a port scan of the Dell 2300 reveals that only port 80 is open, as opposed to the Dell 1184 which has many other ports open. Therefore, the following hack does not apply to the Dell 2300.

Performing the Hack

To begin, fire up your computer and open a DOS prompt or UNIX shell. Telnet to the IP address of your 1184 gateway or router on port 333. By default, this address is 192.168.2.1:

```
C:\>telnet 192.168.2.1 333
```

By executing this command, you will be presented with a login prompt. For a username, enter *root*. For the password, enter the password used by the browser interface (Note that the browser interface uses a default username and password of *admin*). So, if your 1184 is set to the factory default configuration, you can log into the Telnet daemon using the username *root* and the password *admin*. A successful login will look like this:

```
login: root
Password: (the password will not be shown as you type it)
#
```

That's it! You're root and you "own the box", meaning you have complete control of the entire system. Want to learn more about your AP? Use the command:

```
sysconf view
```

Running this command will give you the following output (note that some parameters may vary with different AP firmware versions):

```
Sysconf Reading System Parameters from FLASH...DONE!
current parameter size 4204

+===================================================+
|         System Configuration Table : valid !!!        |
+===================================================+
|                    System Parameters                  |
|   Host Name          :                                |
|   System User ID     : root                           |
|   System Password    : admin                          |
+---------------------------------------------------+
|                   Boot Configuration                  |
|   Boot Method : Auto Boot                             |
|   Vendor Name : Gemtek Taiwan                         |
|   Boot File Name : /home/tftp/vLinux.bin.gz           |
|   TFTP Server IP Address : 192.168.2.239              |
+---------------------------------------------------+
|            LAN (Ethernet) Configuration               |
```

```
|  Ethernet H/W Address : 00:90:4b:08:30:75           |
|  Ethernet IP Address : 192.168.2.1                  |
|  Ethernet Default Gateway : 0.0.0.0                 |
|  Ethernet Subnet Mask : 255.255.255.0               |
+-----------------------------------------------------+


+=====================================================+
|                     WAN Setup                       |
+-----------------------------------------------------+
|  No Connection Type Selected                        |
+-----------------------------------------------------+
|                  PPP Configuration                  |
|  PPP User Identifier :                              |
|  PPP User Password   :                              |
|  PPP Using PAP Authentification                     |
|  PPP Single Connection                              |
+-----------------------------------------------------+
|          IP (Cable Modem) Configuration             |
|  MAC H/W Address   : 00:90:4b:08:30:75              |
|  IP Address        : 0.0.0.0                        |
|  Default Gateway   : 0.0.0.0                        |
|  Subnet Mask       : 0.0.0.0                        |
|  Using DHCP for WAN Port                            |
+-----------------------------------------------------+


+=====================================================+
|             DHCP Server Configuration               |
|    Invalid DHCP Configuration                       |
+-----------------------------------------------------+
|              DNS Server Configuration               |
|  Domain Name Server      :                          |
|  Primary DNS Server IP   : 0.0.0.0                  |
+-----------------------------------------------------+
|        NAT Virtual Server Configuration             |
|  Virtual   Server : 0.0.0.0                         |
|  Virtual   Server : 0.0.0.0                         |
|  Virtual   Server : 0.0.0.0                         |
```

```
|  Virtual   Server : 0.0.0.0                              |
|  Virtual   Server : 0.0.0.0                              |
|  Virtual   Server : 0.0.0.0                              |
|  Virtual   Server : 0.0.0.0                              |
```

If you wish, you can change these settings with the command:

```
sysconf set
```

Next, let's take a look at the file system:

```
# ls -la
drwxr-xr-x  1 root       root          32  Jan 01 1970  .
drwxr-xr-x  1 root       root          32  Jan 01 1970  ..
drwxr-xr-x  1 root       root          32  Jan 01 1970  bin
drwxr-xr-x  1 root       root          32  Jan 01 1970  cgi-bin
drwxr-xr-x  1 root       root          32  Jan 01 1970  dev
drwxr-xr-x  1 root       root          32  Jan 01 1970  etc
drwxr-xr-x  1 root       root          32  Jan 01 1970  home
drwxr-xr-x  1 root       root          32  Jan 01 1970  images
-rw-r--r--  1 root       root         429  Jan 01 1970  index.html
drwxr-xr-x  1 root       root          32  Jan 01 1970  lib
drwxr-xr-x  1 root       root          32  Jan 01 1970  mnt
dr-xr-xr-x 25 root       root           0  Jan 01 2000  proc
drwxr-xr-x  1 root       root          32  Jan 01 1970  tmp
drwxr-xr-x  1 root       root          32  Jan 01 1970  usr
drw-rw-rw-  3 root       root        1024  Nov 14 23:18 var
```

Running the command...

```
cat /etc/hosts
```

...reveals the following information:

```
127.0.0.1                vLinux/Vitals_System_Inc.
```

It appears that this access point is running the vLinux distribution. More information about vLinux can also be found at www.onsoftwarei.com/product/prod_vlinux.htm. To continue with our experimentation, let's take a look at the boot-up script by typing the following command:

```
cat /etc/rc
```

This will result in the following output:

```
# set default host name
#hostname Wireless_Broadband_Router
# expand /dev/ram0 to ext2 file system
echo "Expand RAM file system image"
/etc/expand /etc/ramfs2048.img /dev/ram0
/etc/expand /etc/ramfs.img /dev/ram1
# mount filesystem and make system directory
echo "Mount filesystem and make system director
mount -n -t proc /proc /proc
mount -n -t ext2 /dev/ram0 /var
mount -n -t ext2 /dev/ram1 /etc/config
chmod 666 /var
chmod 755 /etc
mkdir /var/run
#must be execute before sysconf
checkisp &
reset &
# setup system network
#sysconf config
#sh /etc/bridge
#johnny add
rserver &
reaim &
#server&
#johnny end
#added by tom for start up thttpd
#/bin/thttpd -r -d / &
#/cgi-bin/sysconf.cgi &
```

Did you notice the *#johnny add, #johnny end,* and *#added by tom for start up thttpd* lines? Who are Johnny and Tom? Most likely, they are engineers or developers who helped design the Dell AP or the vLinux distribution. Next, type **ps** to see the list of running processes, which will show something like:

```
 PID PORT STAT SIZE SHARED %CPU COMMAND
   1       S     0K     0K  0.0 init
   2       S     0K     0K  0.0 kflushd
```

```
     3        S        OK     OK   0.0  kupdate
     4        S        OK     OK   0.0  kswapd
    16        S        OK     OK   0.0  checkisp
    17        S        OK     OK   0.0  reset
    18        S        OK     OK   0.0  rserver
    19        R        OK     OK   0.2  reaim
    21   S0   S        OK     OK   0.0  /bin/login
    22        S        OK     OK   0.0  /bin/inetd
    23        S        OK     OK   0.0  /bin/thttpd
    24        S        OK     OK   0.0  /cgi-bin/sysconf.cgi
    61        S        OK     OK   0.0  dproxy
    69        S        OK     OK   0.0  dhcpd
    79        S        OK     OK   0.0  dhclient
  4446        S        OK     OK   0.7  /bin/telnetd
  4447   p0   S        OK     OK   0.0  sh
  4476        S        OK     OK   0.0  /bin/dporxy
  4499   p0   R        OK     OK   0.0  ps
```

Here, you can see the Web server (/bin/thttpd running as process ID 23) and our Telnet daemon (/bin/telnetd running as process ID 4446).

Want to learn more about the operating system version? Type the following:

```
cat /proc/version
```

This will show something similar to:

```
Linux version 2.2.14-v1.9 (root@localhost.localdomain) (gcc version 2.9-vLinux-armtool-
0523) #5357 Sat Jan 25 17:39:42 CST 2003
```

As you can see, the system once again identifies itself as a Linux distribution. What kind of CPU, you wonder? Type this:

```
cat /proc/cpuinfo
```

You will see something like the following:

```
Processor     : S3C4510/SEC arm7tdmi rev 0
BogoMips      : 44.24
Hardware      : <NULL>
```

Now, would you like to play around with the HTML files that are used in the browser interface? Take a look at the files in /home/httpd. Need to see statistics on frequently used objects in the kernel? Try *cat /proc/slabinfo*. How about current memory usage details? Try *cat /proc/meminfo*. You can even play around with IPChains, ping, gzip, ifconfig, reboot, and other utilities. Have fun and explore! The possibilities are endless for experimenting with your Linux-based Dell 1184 Access Point.

Under the Hood: How the Hack Works

Similar to the USR 2450 Access Point, the Dell 1184 hardware is a single-board computer. The Dell Access Point was designed to run an embedded version of the Linux operating system and all that was needed was to Telnet right into its open arms through port 333. It doesn't require any special tools or reprogramming —it's ready to go, straight out of the box, giving you an easy path into the exciting world of hardware hacking!

Summary

In this chapter, we showed three hardware hacks for wireless networking products. In our first hack, we modified a D-Link DWL-650 wireless NIC to add an external antenna. Most consumer-grade cards do not provide an external antenna connection. Those that do are generally more expensive. However, the D-Link card can be modified to give it support for an external antenna with relative ease. In our hack, we snipped off the end of a Thinnet cable and soldered its BNC connector to the available leads on the D-Link card's PCB. By looking at the PCB, it appears as if the D-Link card has support for an external antenna, but it was never implemented.

In our next hack, we explored OpenAP, an open-source Linux distribution from Instant802. The OpenAP software allows you to reprogram certain brands of off-the-shelf access points with a fully functioning Linux operating system. In our hack, we used a U.S. Robotics USR 2450 AP. The USR 2450 has a special jumper on the motherboard that, when shorted, will cause the AP to boot from an SRAM card if one is inserted into the PCMCIA slot. By removing the wireless NIC from the PCMCIA slot and replacing it with a preprogrammed SRAM card containing a OpenAP image file, we can "reflash" the AP's on-board Flash memory. Then we can remove the SRAM card, replace the wireless NIC, and reboot. Voilá! We now have a Linux machine running on an access point.

In our final hack, we explored the inner workings of the Dell 1184 Access Point. The Dell 1184 contains an embedded Linux distribution. No special tools or reprogramming is necessary and we can simply Telnet to the device on port 333 and gain complete access.

Additional Resources and Other Hacks

This section lists a number of interesting Web sites and other wireless-related hardware hacks. If you're interested in learning more about the wireless hacking community or just wireless technologies in general, follow these links.

User Groups

- **San Diego Wireless Users Group** www.sdwug.org
- **Bay Area Wireless Users Group** www.bawug.org

- **Southern California Wireless Users Group** www.socalwug.org/
- **Orange County Wireless Users Group** www.occalwug.org/
- **NYC Wireless** www.nycwireless.net
- **Seattle Wireless** www.seattlewireless.net
- **Personal Telco** www.personaltelco.net
- **Free Networks** www.freenetworks.org
- **Airshare** www.airshare.org
- **Other User Groups** www.wirelessanarchy.com/#Community%20Groups

Research and Articles

- **William Arbaugh, Wireless Research Web Page**, www.cs.umd.edu/~waa/wireless.html
- **Tim Newsham, 802.11 Wireless LAN Web Page**, www.lava.net/~newsham/wlan
- **N. Borisov, I. Goldberg, D. Wagner, (In)Security of the WEP Algorithm Web Page**, www.isaac.cs.berkeley.edu/isaac/wep-faq.html
- **P. Shipley, "Open WLANs: The Early Results of War Driving," 2001**, www.dis.org/filez/openlans.pdf
- **S. Fluhrer, I. Mantin, A. Shamir, "Weaknesses in the Key Scheduling Algorithm of RC4," Aug 2001**, www.wisdom.weizmann.ac.il/~itsik/RC4/Papers/Rc4_ksa.ps
- **IEEE Standards Wireless Zone Web Page** standards.ieee.org/wireless
- **IEEE 802.11 Working Group Web Page** grouper.ieee.org/groups/802/11
- **WarDriving.com Web Page** www.wardriving.com

Products and Tools

- **Airsnort**, 64/128-bit WEP key cracker based on flaws in RC4 Key Scheduling Algorithm, airsnort.sourceforge.net
- **Network Stumbler** www.stumbler.net
- **MacStumbler (OS X)** www.macstumbler.com
- **bsd-airtools (*BSD)** www.dachb0den.com/projects/bsd-airtools.html
- **NetChaser (Palm OS)** www.bitsnbolts.com/netchaser.html
- **KisMAC (OS X)** www.binaervarianz.de/projekte/programmieren/kismac

- **Pocket Warrior (Pocket PC)** www.dataworm.net/pocketwarrior/index.html

- **Nice listing of assorted wireless tools** www.networkintrusion.co.uk/wireless.htm

- **Ethereal** Packet capturing tool, www.ethereal.com

- **prismdump** Retrieve raw 802.11 frames with Prism II-based wireless NICs, developer.axis.com/download/tools

- **WEPCrack** 64/128-bit WEP key cracker based on flaws in RC4 Key Scheduling Algorithm wepcrack.sourceforge.net

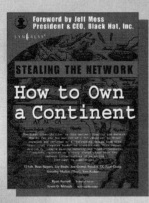

"This novel is scarily realistic."
—James A. Whittaker, Ph.D., Chief Scientist and Founder of Security Innovation

The Mezonic Agenda: Hacking the Presidency

Dr. Herbert H. Thompson and Spyros Nomikos

The Mezonic Agenda: Hacking the Presidency is the first Cyber-Thriller that allows the reader to "hack along" with both the heroes and villains of this fictional narrative using the accompanying CD containing real, working versions of all the applications described and exploited in the fictional narrative of the book. The Mezonic Agenda deals with some of the most pressing topics in technology and computer security today including: reverse engineering, cryptography, buffer overflows, and steganography. The book tells the tale of criminal hackers attempting to compromise the results of a presidential election for their own gain.

ISBN: 1-931836-83-3

Price: $34.95 U.S. $50.95 CAN

Stealing the Network: How to "Own the Box"

Ryan Russell, FX, Joe Grand, and Ken Pfiel

Stealing the Network: How to Own the Box is NOT intended to be an "install, configure, update, troubleshoot, and defend book." It is also NOT another one of the countless Hacker books out there now by our competition. So, what IS it? *Stealing the Network: How to Own the Box* is an edgy, provocative, attack-oriented series of chapters written in a first hand, conversational style. World-renowned network security personalities present a series of chapters written from the point of an attacker gaining access to a system. This book portrays the street fighting tactics used to attack networks.

ISBN: 1-931836-87-6

Price: $49.95 USA $69.95 CAN

SYNGRESS®

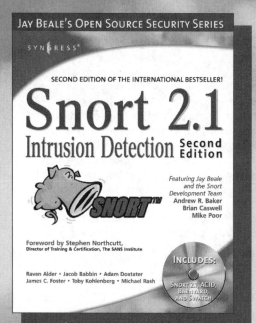

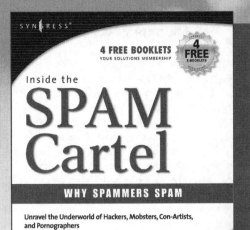

Inside the SPAM Cartel

Spammer X

Authored by a former spammer, this is a methodical, technically explicit expose of the inner workings of the SPAM economy. Readers will be shocked by the sophistication and sheer size of this underworld. "Inside the Spam Cartel" is a great read for people with even a casual interest in cyber-crime. In addition, it includes a level of technical detail that will clearly attract its core audience of technology junkies and security professionals.

ISBN: 1932266-86-0

Price: $49.95 US   72.95 CAN

Penetration Testing with Google Hacks

Johnny Long,
Foreword by Ed Skoudis

Google, the most popular search engine worldwide, provides web surfers with an easy-to-use guide to the Internet, with web and image searches, language translation, and a range of features that make web navigation simple enough for even the novice user. What many users don't realize is that the deceptively simple components that make Google so easy to use are the same features that generously unlock security flaws for the malicious hacker. Vulnerabilities in website security can be discovered through Google hacking, techniques applied to the search engine by computer criminals, identity thieves, and even terrorists to uncover secure information. This book beats Google hackers to the punch, equipping web administrators with penetration testing applications to ensure their site is invulnerable to a hacker's search.

ISBN: 1-931836-36-1

Price: $49.95 USA $65.95 CAN

SYNGRESS®

Syngress: *The Definition of a Serious Security Library*

Syn·gress (sin-gres): *noun, sing.* Freedom from risk or danger; safety. See *security*.